Fabio Neves de Miranda

Teaching and learning strategies in distance education

Fabio Neves de Miranda

Teaching and learning strategies in distance education

An analysis of the importance of teaching-learning strategies in distance education in the digital age

ScienciaScripts

Imprint

Cover image: www.ingimage.com

This book is a translation from the original published under ISBN 978-3-330-76916-8.

Publisher:
Sciencia Scripts
is a trademark of
Dodo Books Indian Ocean Ltd. and OmniScriptum S.R.L publishing group

120 High Road, East Finchley, London, N2 9ED, United Kingdom
Str. Armeneasca 28/1, office 1, Chisinau MD-2012, Republic of Moldova, Europe
Managing Directors: Ieva Konstantinova, Victoria Ursu
info@omniscriptum.com

Printed at: see last page
ISBN: 978-620-8-60444-8

SUMMARY

I dedicate this work to my precious, virtuous, patient and wise wife Adriane Miranda, who is always there for the struggles of life and for my personal, professional and family victories. May this work serve as an example that hard work and dedication, in the end, is the fruit of four "hands": mine and yours!

THANKS

Because from Him, through Him and for Him are all things. I thank God for giving me the wisdom to tread the paths of research, investigation, science, technology, education and distance learning, as well as the challenges I have overcome, the victories I have won and the hope for a better future.

I would also like to thank my family. My father Gilberto Miranda for always believing in his son's potential. My mother, Suely Miranda, who has always supported me and cheered me on in my victories and supported me in my difficulties. My dear daughter Ester Miranda. My great motivation.

I would like to thank Prof. Dr. Adriana Maria Tonini for being a highly dedicated, caring and friendly advisor. Adriana Maria Tonini for being a highly dedicated, caring, friendly, brilliant advisor, and above all a human being who really shares what she knows, because she knows it. Thank you very much for the joy of your guidance.

To my colleagues at CEFET who walked this path with me and who supported me so many times.

To friends from the FORQUAP and AVACEFETMG research groups, to friends from e-Tec CEFET-MG.

I would like to thank the e-Tec teachers and their respective course coordinators and the general coordinator, Professor Dr. José Wilson da Costa, who were essential to this research.

SUMMARY

The aim of this dissertation was to investigate the pedagogical strategies used in the teaching-learning process of the three distance technical courses offered by e-Tec CEFET-MG: Environment, Electronics and Internet Computing. The theoretical contribution was based on the exponents of cyberculture; distance learning teaching; pedagogical strategies (*Chats*, Forums, Learning Objects, Simulators, External Materials, Interactive Activities, *Webquests*, Animations, *Wiki* (collaborative texts, among others) of teaching and learning harmonized with the pedagogical strategies used by e-Tec CEFET-MG. With regard to the methodological procedures used, this research opted for a qualitative approach, using semi-structured interviews and documentary research to collect data. With the object of the research being pedagogical strategies in the three e-Tec technical courses in the distance modality and the teachers of the three courses as subjects, it was possible to analyze the following categories in the interviews: training and teaching experience; teaching-learning in distance education and pedagogical strategies in distance education with their respective audio transcriptions. Two teachers from each technical course were interviewed in order to obtain information on their respective activities based on the categories mentioned. The results indicated the need for a physical, technological and structural structure for the production of lessons; greater closeness between the coordinators and the teachers; the creation of video lessons to train teachers in the pedagogical strategies of the subjects; a constant effort on the part of the teachers to make the content interdisciplinary; better interaction on the part of the teachers in relation to the methodological proposal contained in the Pedagogical Political Project; real involvement on the part of the Institution's management in relation to CEFETMG's e-Tec Project; constant reinforcement from teachers to students regarding self-discipline, self-organization when it comes to distance education; more time and dedication on the part of teachers to better prepare the pedagogical strategies used in the courses; re-evaluation on the part of the pedagogical team regarding the efficient use of the *Chat* resource in distance education classes. In this sense, the research points to the need to develop methodologies that enable, encourage and train DE teachers to use the various pedagogical strategies that can contribute to the teaching-learning process of students who use this modality to obtain the desired training.

Keywords: Distance Education, Pedagogical Strategies in Distance Education, Teaching and Learning.

1 INTRODUCTION

With the increase in user participation and, consequently, interaction, from the beginning of the 2000s onwards, the *web* began to be seen and used in a different way to what it had been until then, gradually ceasing to be a repository of information and becoming a place for building collective intelligence, as Levy (1996) points out.

The information contained on the web is now fed by people from all over the world. This gave rise to the term Web 2.0, first coined by Tim O'Reilly in 2005, which means an interactive web.

Web 2.0 is the business revolution in the computer industry caused by the shift to the Internet as a platform, and an attempt to understand the rules for success on this new platform (O'REILLY, 2005, p. 78).

Technological changes are occurring and transforming rapidly, generating a significant demand for education that can meet the particular needs of a society, which is a problem for teachers. According to Moran (2007), teachers need to build the necessary skills to work in online education, for example: content mastery, mastery of Information and Communication Technology (ICT) tools and the virtual learning environment, and pedagogical mastery of the online education modality.

According to Moore and Kearsley (2008), distance education offers students flexibility in terms of time and location, since the use of the Web as a medium for distance education offers an open and effective education system that facilitates the selection and appropriation of information and knowledge, reducing the barriers of distance, without losing sight of the educational goals it sets out to achieve.

Seeing this emerging context, the Ministry of Education proposed the creation of the e-Tec Brazil Network as a policy for expanding vocational education, The Ministry of Education, seeing this emerging context, proposed the creation of the e-Tec Brazil Network as a policy to expand vocational education, through the coordination of the Department of Distance Education and the Department of Vocational and Technological Education, which issued Notice No. 01/2007/SEED/SETEC/MEC on April 27, 2007, in which CEFETMG participated and approved three technical courses (REDE e-Tec BRASIL, 2012).

CEFET-MG, in the area of Technological Education, has a type of verticalized teaching offer (from high *school* to post-graduate studies), staff and technological resources that are potentially suitable for excellence in teaching and research in distance education, making it possible to contemplate the inseparability between teaching, research and extension in this form of education as well.

The educational nature of distance learning at CEFET-MG proposes actions that should be based on:

> Because of its inclusive and historic nature, its pedagogical excellence and its social and heuristic nature. Thus, actions in the area should respect the institution's history, implement a good quality teaching policy to the detriment of educational productivism, be democratic, involve regular reflection and problematization through a continuous evaluation system appropriate to the characteristics of this type of education, as well as encouraging research in the area (REDE eTec BRASIL, 2012).

Thus, among e-Tec's goals are: a) to provide high school technical professional education in the distance mode; b) to democratize the offer of free public technical education through distance high school professional training; c) offer high school technical professional education courses in the Environment; d) plan and manage the information technologies of the e-Tec Brasil Program, of the MEC's Distance Education (SEED) and Professional and Technological Education (SETEC) Secretariats (REDE e-Tec BRASIL, 2012).

Faced with this scenario of democratization of technical education in distance learning, questions arise regarding the teaching-learning process in the three technical courses, namely: Electronics, Environment - MA and Computer Science for the Internet, which need to be investigated and which will be the focus of this dissertation.

1.1. Research problem

The Web 2.0 scenario stands out for its combination of social aspects with *online* digital technology. This communication dynamic must be able to connect people, enable sharing and boost collaboration. In this context, Distance Education has been proposed as a modality that really does enhance educational objectives.

*For Castells (2003) and Mill (2006), the media age or information age has an unquestionable impact on education, especially in distance learning. Digital technologies have brought with them new possibilities for communication and interaction.

In order for distance education to achieve its educational purpose of providing quality teaching and consistent learning, there has been intense debate in educational circles. Nascimento and Carnielli (2007) report that distance education:

> It's education and it has to be of the same quality as face-to-face education. Quality distance education is that which helps students learn, just like face-to-face teaching. This is not measured by the number of students involved, but by the seriousness and coherence of the pedagogical project, the quality of the managers, educators and mediators, and also by the development of the student (NASCIMENTO and CARNIELLI, 2007, p. 20).

For Carline and Tarcia (2010), using distance education doesn't mean abandoning learning;

on the contrary, it should create the conditions for students to develop their skills and abilities despite the formal content, as well as forming them as people and citizens as they read, express their reflective and critical thinking in debates, use the virtual environment responsibly and collaboratively, develop study habits and organize work routines.

The authors Carline and Tarcia (2010) also point out that it is educators who guide this educational process. It is important to remember that education is a process; the formation of man is a process. In this sense, it is essential that teachers guide their students to use technologies, in this case through distance learning, in a creative, supportive, collaborative, critical and healthy way, based on pedagogical criteria and parameters that the teachers themselves have helped to build.

In the Brazilian context, Decrees such as No. 5.622 of December 19, 2005, which regulates article 80 of Law No. 9.394/96, reinforce the concern that Distance Education must have with pedagogical didactic mediation in the teaching-learning processes, together with the development of educational activities aimed at quality learning (BRASIL, 2005).

Faced with these educational challenges in today's world, such as distance learning, it is essential to investigate the pedagogical strategies that teachers have used in order to enhance the teaching-learning process.

In order to help this way of thinking and doing education, the aim of this dissertation is to investigate how teachers work in distance learning in e-Tec CEFET-MG technical courses by using pedagogical strategies to improve the teaching-learning process.

More specifically, the following problem question arises: What pedagogical strategies are used in the teaching-learning process by teachers of e-Tec CEFETMG distance learning technical courses?

1.2. Justification

Justifying research means highlighting the importance; the relevance of the study to science. It implies highlighting the contribution that the study and its results could make to the scientific community. It can be translated into the opportunities that the research can provide by allowing people to experience and practice topics of interest and the practical advantages and possible applications that the study will make possible (MICHEL, 2009).

In this sense, the offer of distance learning high school technical courses is an alternative to society's current productive and business arrangements, given the growing demands imposed by the productive sector on society and the transformation in the world's attitude towards the demands for training and continued professional qualification. All this as a way of promoting productive and social development, taking into account the applications of scientific knowledge. In this way, the growing demand for professional qualification can be met simultaneously, covering several geographically distinct regions.

In the field of education, the challenge is distance learning and the role of the technical teacher. In the face of this, society is experiencing a historic moment that demands new roles and new skills from professionals - never before demanded with such intensity and with the aim of acting effectively in the intelligence of social and productive processes. Teachers, as professionals, are no exception to these demands (REHEN, 2009).

The emerging scenario itself becomes a lever to start justifying the research proposed in this dissertation. For what reason?

Rehen (2009) points out that the demands on both professionals and the educational field have been shaped by the social, political, economic, productive, cultural and technological transformations in society, combined with the speed with which they occur. Impermanence, transience and fluidity are hallmarks of contemporaneity, imposing new capacities to deal with them in an innovative way.

Faced with this scenario, it is necessary to review and develop pedagogical relationships in order to propose a new way of constructing knowledge, since in distance education the contact between teacher and student is indirect but integrated.

The methodology used to develop distance learning courses means that the content is dealt with and organized in such a way as to make it easier for students to understand without the physical presence of a teacher. It also means that this modality can be applied with or without information and communication technologies - ICT. However, the use of printed material - a mediating technology used mainly during the first generation of distance education - is indispensable (BELLONI, 2009).

In this sense, distance learning is not a technological fad, but the evolution of a long educational process" (PEDROSO, 2006, p. 45). Its strategies need to be constantly reviewed in order to achieve the desired objectives within the teaching-learning process.

In this context, the teacher's didactic, educational and not just technical preparation becomes fundamental. From a formal point of view, for a teacher to work in an institution that offers technical education, such as CEFET-MG, or higher education, they generally need to have a degree, and in some cases a postgraduate degree, which can be *latu-sensu* (specialization) or *strictu-sensu* (master's, doctorate).

In general, teachers are not required to have any didactic training in learning theories, teaching tools and techniques, the teaching-learning process and assessment, except during the selection process (MOREIRA, 2010, p.4).

Teachers have Master's and Doctorate degrees in their specific technical areas, but in most cases they have not undergone any systematic didactic-pedagogical training, except for those who work and have a degree. The justification for this is that teachers don't need didactic training so much as mastery of technical knowledge According to Gil (2012).

According to the author, these justifications were accepted for a long time, mainly because of the elitist nature of education in Brazil. However, the current situation shows that, every day, many students have access to professional, technical and university education. As a result, courses are becoming more specific, challenging teachers to acquire not only technical competence in their field, but also didactic and pedagogical competence.

The result of all this is that, in general, teachers who find it difficult to teach try to get around the difficulties by carefully, responsibly repeating what is in the textbook. They do this seriously, but are often frustrated because they don't understand why the students don't learn what they're taught. In some cases, the content "is learned without meaning, purely memoristically" (MOREIRA, 2010, p. 4).

The relevance of this research is due to some essential factors regarding professional technical education in Brazil. According to Rehen (2009), there is a certain lack of pedagogical training for those who work in professional technical education, based on the false belief that it is enough to know how to do or know a certain profession or to master only the content of the discipline in question.

Another important aspect in terms of the relevance that justifies this research is the fact that society is facing rapid changes in the world of work, in which learning environments should be enhancers and creators of opportunities. In the contemporary world, the idea of learning is taking on new contours; a new profile of teacher, student and teaching-learning strategies. Teachers as well as students must be willing to align their profiles with new strategic and pedagogical concepts - which require new actors, builders of knowledge, researchers, questioners, creative critics, willing to learn continuously and reform their thinking.

Mill (2012) points out that there are pedagogical principles for virtual education:

During their education, students are primarily responsible for their own construction of knowledge. However, in this process they are the targets of a teaching action (TA), which accompanies them during their studies of didactic-pedagogical materials (DM). These materials are generally organized in multiple media (print, virtual, audiovisual, digital, *web conferencing*, portable, etc.), with a view to redundancy - in other words, the same content can and should be organized in two or more media to cater for different learning styles (MILL, 2012,

p. 37).

In distance education, pedagogical mediation takes place through texts and other resources made available to the student. The treatment of this content and its forms of expression are important to investigate in order to make the educational act effective from the perspective of an education conceived with participation, collaboration, expressiveness and creativity.

In cyberculture, the content and communication interfaces of the Internet are being considered as potential spaces for the collective and collaborative construction of knowledge.

For this to actually happen, Alves (2009) points out that it is necessary:

> The use of teaching strategies to help this process. In this sense, there is a need for research into ways of creating more interactive teaching situations (ALVES, 2009, p. 12).

Teaching is characterized by the permanent challenge for education professionals to establish interpersonal relationships with students, so that the teaching-learning process is articulated and the methods used meet the objectives set.

In this way, students expect the teachers of specific subjects to be outstanding, looking up to them as professional role models and expecting them to pass on the knowledge and methods necessary for them to excel in the future job market. The way in which the teacher plans his or her virtual classroom activities is a determining factor in whether the group of students reacts with greater or lesser interest and contributes to the way the lesson runs, according to Moreira (2010).

The fundamental justification that motivated this research was the possibility of investigating the extent to which teachers are using pedagogical strategies judiciously.

This justifies investigating pedagogical strategies, as they can help to improve the teaching-learning process in the e-Tec CEFET-MG distance learning technical courses.

1.3. Objectives

The aim of this dissertation is to investigate the pedagogical strategies used in the teaching-learning process of the three distance learning technical courses: Environment, Electronics and Internet Computing at e-Tec CEFET-MG.

In view of this objective, the following specific objectives stand out:

a) to identify the pedagogical strategies proposed for distance education in the Political Pedagogical Project (PPP) of the eTec technical courses;

b) to identify with teachers the pedagogical strategies used in the teaching-learning process of the e-Tec CEFET-MG distance learning technical courses;

c) to analyze the use of pedagogical strategies that promote learning in the distance modality of technical courses based on the e-Tec PPP and the theoretical references researched.

The outline of the theme, problem, objectives and specific objectives of this dissertation are set out in Table 1.

Chart 1 - Outline: theme, problem and objectives

<table>
<tr><th rowspan="2">Theme</th><th rowspan="2">Problem</th><th colspan="3">Objectives</th></tr>
<tr><th>General</th><th colspan="2">Specifics</th></tr>
<tr><td rowspan="3">Pedagogical teaching-learning strategies in the distance mode</td><td rowspan="3">What pedagogical strategies are used in the teaching-learning process by teachers of eTec CEFETMG distance learning technical courses?</td><td rowspan="3">To investigate the pedagogical strategies used in the teaching-learning process of the technical of the three technical courses: Environment, Electronics and Internet Computing at e-Tec CEFET-MG</td><td>Identify</td><td>in the Political Pedagogical Project - PPP of the e-Tec technical courses the pedagogical strategies proposed for Distance Education.</td></tr>
<tr><td>Identify</td><td>the pedagogical strategies used in the teaching-learning process of the e-Tec CEFET-MG distance learning technical courses.</td></tr>
<tr><td>Analyze</td><td>the use of pedagogical strategies that promote learning in the distance modality of technical courses based on the e-Tec PPP and the theoretical references researched.</td></tr>
</table>

Source: Prepared by the author

1.4. Dissertation Structure

Using the ABNT NBR 14724:2011 standard (ABNT, 2011), this dissertation is organized as follows: introduction, literature review, methodology and concluding remarks, in six chapters, this being the first. Table 2 shows the sections with the main theoretical references used in the literature review.

Table 2 - Structure of sections and main authors

SECTIONS AND LITERATURE REVIEW	
Introduction	O'Reilly (2005); Moore & Keasley (2008); Castells (2003); Rede e-Tec brasil (2007);
Distance education in cyberculture	Levy (2003); Mill (2001); Garcia (2012); Mattar (2008);
Teaching in distance education	Mill (2012); Silva (2012); Tardif & Lessard (2005); Pimenta and Anastasiou (2010); Morris and Finnegan (2005); Behar (2013); Coronado (2009);
Pedagogical strategies for Teaching and learning	Libâneo (1994); Candau (1984); Petruci & Batiston (2006); Dougiamas and Taylor (2002); Mazzioni (2009); Almeida (2003); Peters (2003).
Pedagogical strategies at e-Tec CEFETMG	Silva (2010); Dodge (2013); PPP e-Tec (2008); Costa (2004);
Research into the pedagogical strategies of e-Tec	Mazotti (2001); Michel (2009); Trivinos (2009); Barros and Lehfeld (2009)

Source: Prepared by the author

The second chapter deals with distance education in the context of cyberculture and CEFET-MG. In the third chapter, the role of the distance learning teacher will be investigated. Chapter four discusses the pedagogical strategies needed to enhance the teaching-learning process.

The fifth chapter details the methodology used in the research. It also describes the methodological perspective adopted and then the methodological procedures used (population, instruments, data collection and processing) as well as the analysis of the research data. The results found were described, as well as the correlation between these results according to the theoretical frameworks investigated. The results found and their significance for a better understanding of pedagogical strategies by e-Tec CEFETMG teachers were discussed. The recommendations perceived in this research process were also presented, emphasizing the most relevant aspects derived from the initial studies to the interpretation of the data collected and analyzed, so that contributions can be generated in the academic field.

Chapter six contains the final considerations and proposals for future studies.

2. DISTANCE EDUCATION

In this chapter, Distance Education will be contextualized in the context of Cyberculture, e-Tec CEFET-MG and its Political Pedagogical Project.

2.1. Distance Education in Cyberculture

Cyberculture is the contemporary culture structured by digital technologies that have been promoting new ways of interacting and new learning possibilities mediated by virtual space. Given the specific nature of educational practices, teaching and learning processes can be carried out through virtual learning environments, as stated by Santos (2002).

From Lévy's point of view (1999), cyberculture represents a set of techniques, practices, attitudes, ways of thinking and values that have come into being with cyberspace. It can also be understood as the union of networks and communication resources formed by the global interconnection of computers. It has made it possible to access a computer's resources remotely, to exchange digital files in a simplified way, to send synchronous or asynchronous messages, to hold electronic conferences in real time (video conferencing), to establish electronic business and commerce, to transmit video/sound on demand and many other possibilities that are emerging all the time in today's world.

In the knowledge society, the change in teaching practice is amplified when teachers are faced with a new category of knowledge: digital. For Levy (1993), there are three eras: oral, written and digital, with the latter presenting itself as a significant speed of communication. The author also stresses that recognizing the digital age as a new form of knowledge category does not mean completely abandoning the path trodden by oral and written languages, nor does it mean giving a highly mystical tone to the unbridled and indiscriminate use of technologies, but it does mean confronting telematic and electronic resources with efficient criteria as tools that enhance the most significant methodological processes for student learning.

David Harvey (2001) and several other authors are unanimous in stating that society is experiencing an abyssal change in cultural, political and economic practices. Everyone lives in an era in which capital makes use of information and communication technologies, of all the techniques and technologies that structure the new communication networks, to colonize individuals themselves, whether they are interested in it or not.

Mill (2012) corroborates this by stating that:

> These techniques and technologies now enable new forms of control and social cohesion that are increasingly effective, since they are also more enjoyable. Telematics has brought the possibility of capillarizing society

and colonizing the subjectivity of each of its members. This era of media convergence, configured under a new form of capitalist manipulation that is more perverse, although less aggressive towards individuals, is called the Media Age (MILL, 2012, p. 40).

For Thompson (1995), new ways of thinking and living together are being developed in the world of telecommunications and information technology. Relations between men, work and intelligence itself depend, in fact, on the incessant metamorphosis of information devices of all kinds. This is Pierre Levy's argument, when he says that writing, reading, seeing, hearing, creating and learning are all captured by increasingly advanced information technology.

Mattelart (2002), investigating the history of the information society, points out that, little by little, alphabetic writing and the press suffered their first major threat in the 20th century, caused by the era of electronic transmission.

Still on the subject of the information society, Favacho and Mill (2007) emphasize that defining contemporary society is an arduous task since it is not a question of one society, but of multiple contemporary societies, given the number of paradoxes of a philosophical, anthropological, political, technological and scientific nature.

For Lévy (1993), learning is conditioning a new concept of intelligence - collective intelligence:

Collective intelligence should be understood as "an intelligence distributed everywhere, incessantly valued, coordinated in real time, which results in an effective mobilization of competences".

Therefore, valuing the knowledge of each individual becomes fundamental for the community, since there is no one being who possesses all existing knowledge, but rather beings with different degrees of knowledge. In this sense, each individual has their own way of searching for and systematizing information, and there is always an enormous amount of experience to be shared among everyone, a sharing that is characterized as collective (LÉVY, 1993, p.28).

In the systematization of information, computer science has become fundamental. Lévy (1993) also argues that a human/machine interface refers to the set of programs and material devices that enable communication between a computer system and its human users.

Educational technologies are part of this context of sharing, cyberculture, the age of networks and interfaces. Modern educational technology contributes decisively to making the most important educational directions possible. Garcia (2012) lists five of these directions:

a) the extension of educational opportunities to an ever-increasing number of people is made possible thanks to multimedia that reinforce conventional education; b) the rapid outgrowth of knowledge in schools in general is made up for by the ease with which the mass media transmit education - which involves the individual practically from birth to death; c) the use of appropriate tools facilitates the evaluation of all educational activities conveyed by modern technologies, guaranteeing the efficient use of the means in line with the intended ends; d) educational technologies place the individual; e) the concept of good education is no longer seen from the perspective of "learning everything", but at the level of learning that contributes significantly to new learning (GARCIA, 2012, p. 116). 116).

According to Lévy (1999), cyberspace is an inevitable reality in which time and space are less tangible. Distance learning is part of this context when the author argues that:

It is in cyberspace, created from technological articulations in the areas of information technology and telecommunications, that new educational processes take place, especially distance education. These new times and spaces are basic components of what we call cyberculture (LÉVY, 1999, p. 30).

Lévy (1999) explains that there are reforms needed in today's culture. One of them is the concept of work. Work means learning, transmitting knowledge, producing knowledge. Cyberspace supports intellectual technologies that amplify, externalize and modify numerous human cognitive functions: memory (databases, hyperdocuments, digital archives), simulations, perception (digital sensors, telepresence, virtual realities), reasoning (artificial intelligence), among others, favouring new forms of information, new styles of reasoning. In this sense, reforms are needed in education and training. Distance education is one of them because it exploits teaching techniques, including hypermedia, interactive communication networks and all the intellectual technologies of cyberculture. In this way, teachers are facilitators of the collective intelligence of their groups of students rather than just providers of knowledge.

Faced with this scenario of transformations brought about by cyberculture is distance education. Distance Education

is a type of education in which teachers and students are separated, planned by institutions and using various communication technologies (MAIA and MATTAR, 2008, p. 6).

EaD is a modality whose essential characteristic is the proposal to teach and learn without teachers and students needing to be in the same place at the same time. For learning to take place, technologies and tools are used, computer programs, books, internet resources, available in the virtual learning environment (VLE). Interlocution is possible both through technological supports, for synchronous/simultaneous communication (*web conferences,* videoconferencing, chat, chat room) and for asynchronous communication (forums, text editing tools, *e-mails)* according to Mil (2012).

It is within this context of the Middle Ages that the current model of distance education is configured that virtual distance education teachers are inserted. In view of this fact, it is necessary to take a brief, panoramic look at distance education throughout history in the world and in Brazil.

According to Alves (1994), distance education began in the 15th century with the invention of the printing press and Johannes Guttemberg's composition of words using movable characters. The process of setting up distance learning has a long history. The most recent experiments, around the 18th century and greatly expanded from the 19th century onwards, aimed to generate learning for physically distant individuals through correspondence.

According to Mill (2008), at the end of the 19th century, Distance Education (DE) emerged in the United States and Europe as an alternative to meet the demand for professional knowledge from people who lived in places far from the most developed centers, who had not attended school at the right time, or who had experienced failure in this process.

Moore and Kearsley (2008) state that the origins of distance education go back a long way, and the first experiences are recorded in Sweden in 1883. In 1840, there were reports of distance learning in England. Around 1874, private institutions in the USA and Europe offered correspondence courses aimed at teaching subjects and problems related to trades of little academic value. But, like any historical process, the implementation of distance education did not follow the same timeframe in all countries, since it depended on the society and the objectives for which it was intended.

The author also states that the process of evolution of distance education is divided into five generations, identifiable by the advances in the main communication technologies used. In Brazil, distance education is marked by a trajectory of success, despite the existence of some moments of stagnation, caused mainly by the absence of public policies.

Preti (1996) points out that it was in the 1920s that distance education began in Brazil with the founding of Ràdio Sociedade do Rio de Janeiro by a group from the Brazilian Academy of Sciences. Education via radio was the second means of distance transmission of knowledge, preceded only by correspondence. Television for educational purposes was used positively in the initial phase, and various incentives took place in Brazil, especially in the 60s and 70s, when the current conceptions of distance education began to take new directions, because, despite maintaining written materials as a basis, it began to incorporate audio and videocassettes in an articulated and integrated way.

The first piece of legislation to refer to this modality is the Law of Guidelines and Bases of National Education, whose origins date back to 1961. In its reform, ten years later, a specific chapter on Supplementary Education was inserted, stating that this course could be used in classes, or through the use of radio, television, correspondence and other means.

Junior (2011) points out that in 1993, with the creation of the National Distance Education System (SINEAD), the first steps were taken to implement a national distance education policy. In 1995, the Secretariat for Distance Education (SEED) was created within the Ministry of Education and Culture (MEC), seeking to concentrate efforts with the Ministry of Telecommunications and the Brazilian Post and Telegraph Company. In addition to a major incentive for distance education research projects, such as the MEC/SEED PEPEP (Support Program for Distance Education Projects), SEED began to coordinate the programs: TV Escola, ProInfo, among others.

Distance education was strengthened by Law 9.394 of December 1996, when the country was introduced to a new LDB that made distance education possible. Article 1 of Decree 5.622/2005 of Brazilian legislation provides an important definition of distance education:

> Distance education is characterized as an educational modality in which didactic-pedagogical mediation in the teaching-learning processes occurs with the use of information and communication media and technologies, with students and teachers carrying out educational activities in different places or at different times (BRASIL, 2005).

For Junior (2011), the treatment given to distance education by the LDB encouraged many educational institutions to research and implement distance education systems. From the end of the 1990s, distance education began to differentiate itself, creating its own structure that broke with the boundaries of rigid regulations for face-to-face teaching. Distance learning courses continued to grow under Law No. 9.394/1996 in art. 80, which was later regulated by Decrees No. 2.494 and No. 2.561 of 1998, but revoked by Decree No. 5.622/2005, which is still in force.

2.2. Distance Education at CEFET-MG

In the context of Decree No. 5.622/2005, distance education proposals within the scope of technical schools are taking their first steps. In accordance with Decree 6.301 of December 12, 2007, which establishes the creation of the Open Technical School System of Brazil - e-Tec Brasil, with a view to developing professional and technical education in the form of distance learning. With the aim of broadening the offer and democratizing access to free, public high school technical courses in the interior of the country and on the outskirts of metropolitan areas, the Federal Technological Education Center of Minas Gerais - CEFET-MG is proposing a project to offer high school technical courses in distance learning mode, together with municipal and state public education institutions operating as a network, as face-to-face support establishments (poles) to offer high-level technical courses in the

Environment, Planning and Management in Information Technology and Electronics (REDE e-Tec BRASIL, 2012).

The purpose of the e-Tec Brasil program is to provide professional and continuing training for young students enrolled in and graduating from high school, as well as their teachers. It is one of the actions of the Education Development Plan and aims to take technical courses to regions far from technical education institutions and to the outskirts of large Brazilian cities, encouraging young people to complete secondary education (REDE e-Tec BRASIL, 2012).

With this in mind, the Federal Technological Education Center of Minas Gerais - CEFET-MG submitted proposals for technical courses in the Environment, Planning and Management in Information Technology and Electronics to be offered in 2008/2009 at the presidential support centers selected by SEED/SETEC and published in the DOU of July 4, 2008. This project complied with the above-mentioned call for tenders, the result of which was approved in the selection of courses - Part B of the call for tenders - published in DOU section III, in which federal, state or municipal public institutions were invited to take part, state or municipal institutions that provide secondary technical education that present proposals for secondary technical professional education courses concurrent with or subsequent to secondary education, in the form of distance education (REDE e-Tec BRASIL, 2012).

The e-Tec Brasil Network aims to develop professional and technological education in the form of distance learning. Professionalization, including distance learning, must be an element that contributes to young people and adults entering, remaining in and completing secondary education. In this sense, it is seen as a strategy for raising the level of schooling and should be linked to the other actions of the institution itself, strengthening the possibilities of permanence and continuity of studies (REDE e-Tec BRASIL, 2013).

In the e-Tec Political Pedagogical Project at CEFETMG, various guidelines are set out for the teaching-learning process to take place effectively. It is essential to investigate it (with a focus on pedagogical strategies) in order to propose suggestions that could benefit students, teachers and the institution alike.

2.2.1 e-Tec's Political Pedagogical Project

Various authors provide important definitions of the pedagogical, sociological, epistemological and didactic aspects of the Political Pedagogical Project (PPP).

Silva (2000) says that in the field of academic management, in addition to the Institutional

Pedagogical Project (PPI) and the Institutional Development Plan (PDI), there is the Political Pedagogical Project (PPP), which should not just exist as a statement of intent or a bureaucratic project. The PPP must go beyond curriculum reformulation. It is a collective effort, with the participation of teachers and students in a dialogical process. It is also a permanent, growing process, developing a culture of projects in the institution. It is a working tool that shows what is going to be done, when, in what way, by whom, to achieve what results.

For De Rossi (2004), the Political Pedagogical Project (PPP):

> It is the written document and the instrument of articulation between ends and means: it orders, feeds back and modifies all pedagogical activities, with a view to educational objectives. It takes into account what has been instituted (legislation, curricula, content and methods) and also institutes school culture, since it creates objectives, instruments, procedures, ways of acting, values, etc. (DE ROSSI, 2004, pp. 32-33).

Godoy & Murici (2004) understand that the Political-Pedagogical Project covers three dimensions: a) the *Situational* framework, which is the description of the scenarios, i.e. the reality in which the school is inserted; b) the *Doctrinal* framework, which involves the description of the institution's mission, vision and principles; and c) the *Operative* framework, which corresponds to the pedagogical and administrative guidelines that the school must follow in order to achieve its vision, i.e. focused on the Operationalizing procedures of the political pedagogical project. The operative framework is considered to be the curriculum itself, while the situational and doctrinal dimensions provide the foundations for the process and model of curriculum construction.

In this context, e-Tec's Pedagogical Political Project drew up guidelines for the three distance learning courses: Environment, Electronics and IT for Internet.

With regard to the target audience set out in the e-Tec PPP (2008) and in accordance with what is explained in Article 1 of Decree 6301/2007, it is established, within the Ministry of Education, that the Open Technical School System of Brazil - e-Tec Brasil, with a view to developing professional technical education, aims to expand and democratize the offer of free, public, high-level technical courses in the country, through distance education. e-Tec aims to offer distance technical courses to students in the countryside and on the outskirts of metropolitan areas, enabling initial and continuing professional training for students enrolled in and graduating from high school, as well as for the education of young people and adults; contributing to entry, permanence and completion of high school, preparation for work and life in society.

Specifically with regard to the objectives of the e-Tec technical courses, the PPP (2008) points out that: a) the Environment Technical Course, which belongs to the Environment, Health and Safety Technological Axis, aims to train mid-level professionals to collect and interpret environmental information, data and documentation; b) the Information Technology Planning and Management Technical Course, which from 2013 will be called Internet Informatics, belongs to two professional training areas: Informatics and Management, as it has intrinsic characteristics of these sciences; c) the Electronics technical course, which belongs to the Technological Axis Industrial Control and Processes, aims to train mid-level professionals to participate in the development of projects, carry out the installation and maintenance of electronic equipment and systems, among other aspects.

As for the types of courses, hubs, shifts and duration, the e-Tec PPP (2008) explains that the technical courses approved under Notice No. 01/2007/SEED/SETEC/MEC will be offered in the hub municipalities of Alfenas; Almenara, Porteirinha and Timòteo, with 50 vacancies, a workload of 1,300 hours plus an 18-month internship. The selection process involves young people who are attending at least the second year of secondary school.

As a justification, according to the Pedagogical Political Project (2008), the proposal to create a high school technical course offered in the distance education modality aims to be a comprehensive form of education, and proposes to reach all citizens, using a participatory and permanent process, aiming to form in the student a critical awareness of current problems, as well as equipping them with skills for solutions and appropriate planning.

> The methodology used in the development of distance learning courses proposes that the content be treated and organized in such a way as to make it easier for students to understand, even without the physical presence of a teacher. This methodology means that the modality can take place with or without information and communication technologies - ICT, but the use of printed material - the mediating technology used mainly during the first generation of distance education - is essential at the very least (PPP e-Tec, 2008, p. 26).

In addition to educational activities, students will also carry out practical laboratory activities at the hubs in laboratories provided by the institution, as well as compulsory internships. For internships, partnerships will be made with private and public institutions in order to make the internships provided for in the course feasible (PPP, 2008).

As for support, the course will be developed through face-to-face support centers selected under the e-Tec Brazil Program:

> in which enrolled students will receive face-to-face assistance from a team of tutors who will be selected and trained by the research group of the Master's Degree in Technological Education of CEFET-MG's Postgraduate Program, AVACEFET-MG, to perform this function. The e-Tec tutoring process will be face-to-face in the centers and at a distance in the NEaD (Distance Education Center) located in Campus IV of CEFETMG (Projeto Politico Pedagògico e-Tec, 2008, p.32).

The educational infrastructure organized at the educational institution, present at the Distance Learning Center, is complemented by the technology infrastructure at the centers, made up of computer labs with Internet access, teaching labs, videoconferencing rooms and administrative and study spaces, which guarantee the student the necessary conditions to carry out the academic activities of the course.

With regard to strategies, media and didactic units, the Pedagogical Political Project (2008) emphasizes that:

> Their purpose is to help achieve certain levels of learning with greater or lesser ease. Therefore, a methodology will be developed for each one, focusing on the advantages offered by each medium. The development of the didactic units of the curricular components of each module will use these different strategies in conjunction (PPP e-Tec, 2008, p. 33).

Students will have at their disposal one video lesson per didactic unit, which can be watched, paused and replayed as many times as the student deems necessary. The topics of the video lesson can be followed on the AVA - Virtual Learning Environment. A supporting bibliographic text, produced and distributed for each curricular component, will provide details of the main points to be covered. Students will also carry out practical laboratory activities at the hubs in laboratories provided by the institution, as well as the compulsory internships (PPP e-Tec, 2008).

As for the Virtual Learning Environment (VLE) for CEFET-MG's *distance learning* courses, will use the *Modular Object Oriented Distance Learning* VLE - *MOODLE*, which is an LMS (*Learning Management System*) used to manage distance *learning* courses. *Moodle* is *Open Source Software*, which means that you can install, use, modify and even distribute the program at no cost (under the terms of the GNU - *General Public License*). The VLE can be used, without modification, on *Unix, Linux, Windows, Mac OS* and other systems that support the PHP language (PPP E-TEC, 2008).

The teaching materials set out in the e-Tec Pedagogical Political Project (2008) are: a) the course book for the subject; b) the manual for the virtual learning environment used (MOODLE); c) the distance learning course guidelines; d) books and articles used as bibliographic sources for research and subject activities; e) complementary audiovisual materials; f) recommended works.

The political-pedagogical project of the e-Tec courses proposes the use of multiple means (media) to achieve the educational objectives. Thus, in this dissertation, the following are called pedagogical strategies:

> virtual classes; learning objects that will be developed throughout the course; simulators; forums; chat rooms;

connections to external materials; interactive activities; virtual tasks (webquest); modelers; animations; collaborative texts (Wiki) (PPP e-Tec, 2008, p. 106).

According to Gabrielli, Rozenfeld and Soto (2010), in the VLE, it is essential to develop strategies that allow the teacher to be present (or, in certain situations, absent), in order to motivate students to interact, to seek information, to have an autonomous attitude towards their learning process, a characteristic that gives great value to the type of tool and language used.

Due to its characteristics, distance education can help to achieve its social function and CEFET-MG's institutional objectives, since it has the potential to significantly expand the supply of technical education for the professional practice of people who are unable to obtain professional training in their places of origin or residence or to travel to do so.

Against this backdrop of changes at the technical level and in the case of distance learning, the teacher now plays a key role. In this sense, it is essential to highlight aspects of the virtual teacher and their pedagogical mediation, which will be investigated below.

3. TEACHING IN DISTANCE EDUCATION

In order to achieve excellence and pedagogical competence, it is essential to understand teaching, the role of the teacher and the pedagogical strategies that enable the teaching-learning process to be enhanced.

> What all great teachers seem to have in common is a love for their subject, an obvious satisfaction in awakening this love in their students, and an ability to convince them that what they are being taught is terribly serious (EPSTEIN, 1981, p. 13).

Education is inherent in human society. It is present in the home, in the street, in the church, in the media in general and everyone is involved with it, whether to learn, to teach or to learn-and-teach. To know, to do, to be or to live together every day, it is possible to mix life with education. Education is a "natural" process that takes place in human society through the actions of its agents as a whole (BRANDÂO, 1981).

Education is a process of humanizing human society with the aim of making individuals participants in the process of civilization and responsible for carrying it forward. Teaching is part of this. Teaching is an educational practice, in other words, it is a way of intervening in social reality; in this case, through education. Education, as a reflection, portrays and reproduces society, but also projects the society we want (PIMENTA & ANASTASIOU, 2010).

From an educational perspective, teaching has undergone intense changes in recent times, especially in the profile of teachers, from specialists to learning mediators. The current teaching scenario is constantly changing and requires reflection on this domain. Teachers' attitudes are changing

> from the specialist who teaches to the learning professional, who "encourages and motivates the learner; providing a kind of bridge between the learner and their learning, not a static bridge, but a 'rolling' bridge that actively collaborates towards the student's objectives" (MASETTO, 2012, p. 29).

The exercise of teaching, and in particular the art of teaching, is magna. By "magna", didactics from the point of view of Comenius (2011) means to teach in the right way, in order to obtain results; to teach in an easy way without teachers and students getting bored or annoyed; on the contrary, to have great joy in teaching in a solid way, not superficially in any way, but in order to lead to true culture and good habits. Let the bow and stern of didactics be to seek and find a method so that teachers teach less and students can learn more and more (COMENIUS, 2011).

Freire (2011) highlights the difference

between teaching and learning when he says that "there is no teaching without learning, the two explain each other and their subjects, despite the differences that characterize them, are not reduced to the condition of being each other's object. Those who teach learn by teaching and those who learn teach by learning" (FREIRE, 2011, p. 12).

For Masetto (2010), teaching is the mastery of specific knowledge in a given area, which is mediated by a teacher to his or her students. It is an educational action that is constituted in the teaching-learning process, in research, in the management of educational contexts and from the perspective of democratic management. Consequently, teaching work is characterized by processes and practices of cultural production, organization, appropriation of knowledge and dissemination of what is developed in school educational spaces, under certain historical conditions. From this perspective, teachers are defined as subjects in action and interaction with others (teachers/students), producers of scientific knowledge for the reality in which they find themselves.

Tardif (2010) reports that teachers' knowledge is not limited to the transmission of knowledge. Their practice integrates different types of knowledge. In this sense, it is plural knowledge, made up of a more or less coherent amalgam of knowledge from professional training, disciplinary knowledge, curricular knowledge and experience.

As for the subject, being a teacher:

is nothing more than a set of personalized interactions with students in order to get them involved in their own training process and meet their different needs (TARDIF & LESSARD, 2005, p. 267).

For Libâneo (1994), teaching work is the activity that gives unity to the teaching-learning binomial, through the process of active transmission-assimilation of

knowledge, mediating the cognitive relationship between the student and the teaching material.

Malheiros (2012) says that the teacher training process has three fundamental aspects: theoretical training, didactics and practical training. Theoretical training deals with specific knowledge of the subject the teacher will be working in. Didactic training aims to qualify teachers to use and develop appropriate methodologies. In practical training, the aim is to initiate teachers into the academic environment, contextualizing them with the classroom, whether face-to-face or virtual.

For Freire (2011), teaching requires critical reflection on one's own practice. Critical teaching practice implies right thinking, it involves the dynamic, dialectical movement between doing and thinking about doing. The fundamental moment in teacher training is that of critical reflection on their practice. It is by thinking critically about today's or yesterday's practice that the next practice can be improved (FREIRE, 2011).

In this context of practice, Pimenta and Anastasiou (2010) reinforce that educational practice has been commonly identified with the technical dimension of teaching, which characterizes instrumental didactics and involves techniques, didactic materials, class control, curricular innovations, competences and skills of the teacher, according to the prism of effective control of the process.

According to Masetto (2012), teaching has one main objective: student learning. In this sense, *online* education becomes an opportunity to enhance learning.

Online education has generated great opportunities and interactions between teachers, students, content and subjects, and is one of the fundamental elements of the so-called *Web 2.0*.

For Leâo (1999), the use of hypermedia involves technologies that encompass hypertext and multimedia resources, allowing users to navigate through different parts of an application.

For Silva (2012), reflecting on the use of hypermedia is:

"To break with linearized thinking is to understand hypertextuality, observing the most varied situations that arise from the interaction between subject and object in the learning process" (SILVA, 2012, p. 112).

In this learning process, new spaces have emerged, such as virtual environments created through telematics and information technology. The Internet for research, *e-mails*, forums, *chats*, groups, discussion lists, portfolios, websites, *wikis*, videos, teleconferences are new environments that students can navigate to promote their learning. These resources create virtual environments that can complement face-to-face environments or be used in distance learning situations (MASETTO, 2012).

Machado (2008) proposes important teaching actions in the face of information and communication technologies. Among these, weaving meanings and mediating relationships are fundamental. New attitudes are proposed for teachers in *online* education in this way:

master computer and telematics resources in order to be able to use them with students; know how to guide distance activities and work; carry out pedagogical mediation and plan a course with distance activities (MASETTO, 2012, p. 96).

With regard to teaching, Masetto (2010) also points out that teachers today are no longer defined as passers on or transmitters of content, but as mediators. This expression, frequent in pedagogical discourses, characterizes the approaches that oppose the traditional school and is didactically translated into a series of attitudes and didactic procedures. The author also stresses that pedagogical mediation is the attitude, the behavior of the teacher who acts as an encourager or motivator of learning, as a rolling bridge between the learner and learning, highlighting dialogue, the exchange of experiences, debate and the proposition of situations.

Mediation can be seen from two angles, according to D'Avila (2008). Cognitive mediation brings the idea of an intelligent apprehension of the object of knowledge which cannot be immediate, but mediated by the mental prayer of the subject who conceptualizes it, the concept being a human construct. However, for the author, cognitive mediation presupposes didactic mediation, of an external nature. The relationship of knowledge is doubly mediated, i.e. there is a cognitive mediation and another of a didactic nature. The author also stresses that there are in fact two processes of mediation: the one that links the learner subject to the object of knowledge (called cognitive mediation) and the one that links the trainer-teacher to this relationship. Thus, in the relationship with knowledge, there is a double mediation, one of a cognitive nature and the other of a didactic nature. However, there is a certain subordination of didactic mediation to cognitive mediation, which is the learning process, a process of objectification of the real that takes place in the relationship between subject(s) and object(s), in a specific space-time context. Didactic mediation therefore consists of establishing the ideal conditions for activating the learning process.

In distance learning, mediation plays an extremely important role since physical distance always requires different resources, strategies, skills, competencies and attitudes from teachers. With digital technologies, the mediating role of the teacher has taken on a strong impetus, due to the possibilities and also the demands of the configuration of this new "space".

Maia and Mattar (2008) emphasize that teachers in distance education have important facets. They go from speaker to tutor, from lecturer to facilitator, from assessor to mediator. Instead of the *sage on stage*, the *guide on the side*.

Technologies have led to changes in teachers' working methods, generating changes in the way institutions work and in the education system. By using technology in a contextualized way, teachers will help to improve the teaching-learning process, as well as contributing to the formation of a critical, reflective citizen with more educational and professional possibilities. The use of technology does not cancel out the fundamental role of the teacher; on the contrary, it brings about change, because in this case the teacher will take on the role of organizer of collective knowledge, managing and guiding their students to seek out various sources of information and knowledge, as Fidalgo (2011) points out.

Thus, in distance learning, the teacher also acts as a mediator in the process of building knowledge. In this position, some of their functions are: preparing teaching materials; selecting content, which must always be updated; checking the functionality of the planning, making the necessary adjustments; stimulating interaction and, above all, stimulating the presentation, discussion and possible solutions to problems that arise during the teaching-learning process.

Porto (2009) alludes to the paradigm shift in education.

In this case, other skills are needed to develop pedagogical mediation in distance education. The aspects of pedagogical mediation that, according to the author, need to be considered when it comes to distance education are:

> The interaction that must exist between the subjects of the pedagogical relationship and the interactivity between these subjects and the information technology tools available; The organization of time and space for carrying out the teaching-learning project that makes it possible to use the generation in which communication is stored and accessed at different times and in different spaces; The pedagogical application of information and communication technologies, not for their own sake, but which serve to problematize learning situations, generating interactive and cooperative actions, in accordance with the assumptions that have been guiding the production of knowledge and the implementation of learning from a critical-reflective perspective; The concern with the subject as a whole, understanding them from the idea of complexity; among others (PORTO, 2009, p. 53-55). 53-54).

Gutiérrez and Prieto (1994) called the distance learning teacher a "pedagogical advisor". Their role is to make the link between the institution and the student, accompanying the process in order to enrich it with their knowledge and experience. They must be able to communicate well; have a clear concept of learning; master the content well; facilitate the construction of knowledge through reflection, exchange of experiences and information; establish empathetic relationships with the student; seek out philosophies as a basis for their act of educating; and constitute a strong instance of personalization. Among the priority tasks of the "pedagogical advisor" are networking, promoting group meetings and evaluating.

For Niskier (1999), the distance educator brings together the qualities of a planner, pedagogue, communicator and IT technician. They take part in the production of materials, select the most appropriate means for their multiplication, and maintain a permanent evaluation in order to improve the system itself. In this type of teaching, the teacher tries to anticipate possible difficulties, anticipating the students' solutions.

Garcia (2012) states that there must be a consistent relationship between the technologies and the actual role of the teacher. For him, the establishment of a pedagogical relationship, in terms of educational technology, basically depends on the role played by the teacher. It is the teacher who provides the initial elements, who sets the rules of the game, and who sets the content. If teachers don't become aware of the role they play, of the multiple involvements they arouse by simply entering the classroom (whether distance or face-to-face), there will be no point in talking about methods, techniques and so on. The technological relationship runs the risk of becoming a rhetorical expression if teachers don't realize the importance of their role.

In a similar vein, Perez and Castilho (1999) reinforce these competencies with an excellent exposition of the characteristics of the teacher's pedagogical mediation: discussing doubts, questions or problems; providing guidance in technical or knowledge deficiencies; ensuring the dynamics of the learning process; proposing problem situations and challenges; creating an exchange between learning and real society; collaborating to establish connections between acquired knowledge and new concepts, among others.

Mill (2012) carried out a survey on the competencies and profile of teachers in distance education. Among the competencies, some stand out: agile, self-managing, clear, coherent, collaborative, cooperative, cordial, creative, stimulating, manager, mediator, multidisciplinary, participative, proactive, versatile, synergistic, sociable, among others.

In their study, Morris and Finnegan (2005) also propose a typology of teacher roles in distance education: pedagogical, social, administrative and technological. Table 3 lists and summarizes the definitions and activities involved in each role.

Chart 3 - Teacher roles in distance education

Teacher role	Description	Activities
Pedagogical	It involves obligations as an educational facilitator who uses questions and to probe of the of students so that discussions are based on critical concepts, principles and skills.	Provide adequate guidance on tasks; Offering intellectual feedback; Directing comments according to the content and learning objectives; Evaluating students' projects and contributions.
Social	It involves establishing human relationships, developing group cohesion, helping group members to work collectively to achieve a single goal.	To support students in developing collective learning and establishing a sense of community; Offering support in the learning process as well as individual needs
Administrative	It involves organizational, administrative and	Creating discussion forums; Posting

	procedural tasks such as setting dates; timetables, discussion objectives, rules of conduct and decision-making norms.	discussion questions; Answering administrative questions.
Technological	It involves the task of making the use of technology transparent to the students.	Ensure that students are comfortable with the VLE used so that they can focus on the task at hand.

Source: Morris and Finnegan (2005)

The authors Cruz and Barcia (2000) highlight some of the skills that teachers need to develop in order to teach in distance learning: a) planning and organizing courses; b) knowledge of how to encourage collaborative group work; c) mastering questioning strategies; d) possessing in-depth knowledge of the subject content; e) knowing how to involve students and coordinate their activities at a distance; f) possessing a basic knowledge of learning theories; g) mastering knowledge of the field of distance learning.

The works by Cunha (2004), Clark and Peterson (1986) and Coronado (2009) highlight the didactic knowledge needed by teachers in distance education. They are:

> i) the context of pedagogical practice; ii) the learning environment; iii) the planning of activities. The knowledge of the learning environment is fundamental for teachers in distance education, as it provides the teacher with the opportunity to acquire teaching skills, emphasizing learning in order to develop pedagogical possibilities for relating knowledge to social reality, in order to stimulate the students' curiosity and interest. The others are: iv) planning; v) producing activities and teaching materials; vi) conducting the teaching-learning process (CUNHA, 2004; CLARK AND PETERSON, 1990; CORONADO, 2009).

More broadly, from the perspective of distance learning, Belloni (2009) makes a connection between mediatization and strategies:

> Mediatization means designing methodologies and strategies for using teaching/learning materials that maximize the possibilities for autonomous learning. This includes the selection and development of content, the creation of teaching and distance learning methodologies centered on self-directed learning, the selection of the most appropriate media and the production of materials, and the creation and implementation of strategies for using these materials and monitoring the student in order to ensure student interaction with the education system. These strategies must be included in the materials themselves in order to facilitate learning (BELLONI, 2009, p.60).

Sartori and Roesler (2005) point out that in Virtual Learning Environments (VLE), mediation takes place by means of various devices that enable both synchronous and asynchronous communication, making it possible to create various strategies to favor dialogue and the active participation of students. In these environments there is a greater concern with creating moments of interaction and concrete possibilities for collaborative work, with which learning takes place in a participatory way. To do this, teachers can use communication devices to mediate, such as *chats*, forums, *blogs*, *videoblogs* and so on. However, they need to plan how each one will be used and when, preparing themselves to act according to the characteristics and peculiarities of each device.

In order to make the references related to teaching-learning strategies clearer, a brief definition of teaching, teaching process, strategy and the like will be given.

4. TEACHING-LEARNING STRATEGIES

For Libâneo (1994), teaching strategies, techniques and resources are within the context of didactics because it investigates the foundations, conditions and ways of carrying out instruction and teaching. Didactics converts socio-political and pedagogical objectives into teaching objectives, selecting content and methods according to these objectives, establishing links between teaching and learning.

The object of didactics is the teaching process, which includes: the contents of programs and textbooks, the methods and organizational forms of teaching, the activities of the teacher and the guidelines that regulate and guide this process. Libâneo (1994) also provides an important definition of the teaching process:

> The teaching process can be defined as a sequence of activities by the teacher and the students, with a view to assimilating knowledge and developing skills. The purpose of the teaching process is to provide students with the means to actively assimilate knowledge (LIBÂNEO, 1994, p. 54).

Candau (1984) also points out that the object of study of didactics is the teaching-learning process. Every didactic proposal is impregnated, implicitly or explicitly, with a conception of the teaching-learning process. In order to be properly understood, the teaching-learning process needs to be analyzed in such a way that it consistently articulates the human, technical and political-social dimensions.

> The relationship between teaching and learning is not mechanical, but reciprocal. The leading role of the teacher and the activity of the pupil stand out. Teaching aims to stimulate, direct, encourage and drive forward the student's learning process. Teaching has a pedagogical character, i.e. giving a defined direction to the educational process that takes place in the school dimension (LIBÂNEO, 1994, p. 90).

The author also points out that teaching has a two-way nature as it combines the activity of the teacher (teaching) with the activity of the student (learning). The teaching process makes these two poles interact in inseparable moments, when the transmission and active assimilation of knowledge and skills are present.

Therefore, the teacher's main task:

> It's about guaranteeing the didactic unity between teaching and learning through the teaching process. Teaching and learning are two facets of the same process. The teacher plans, directs and controls the teaching process, with a view to stimulating and arousing the students' own activity for learning (LIBÂNEO, 1994, p. 81).

For Candau (1984), the technical dimension refers to the teaching-learning process as an intentional, systematic action that seeks to organize the conditions that best promote learning. Aspects such as instructional objectives, selection of content, teaching strategies, assessment, etc., form the core of its concerns. This is the objective and rational aspect of the teaching-learning process.

And why is it a process? It is a process, according to Libâneo (1994), because it is characterized by the progressive development and transformation of students' intellectual capacities towards the mastery of knowledge and skills and their application. The process aims to achieve objectives in terms of mastering knowledge, skills, habits, attitudes, convictions and the development of students' cognitive capacities.

Teaching, understood as a pedagogical activity, requires three fundamental aspects according to Libâneo (1994): planning, teaching-learning direction and evaluation. Directing teaching and learning requires teachers to: a) know the general principles of learning and how to make them compatible with the subject's own content and methods; b) master teaching methods, procedures, techniques and auxiliary resources; c) have the ability to make teaching content meaningful, real, by referring to the knowledge and experiences they bring to class.

The term "strategy" has been used in a multitude of situations, contexts, realities and with many meanings. The initial definitions are in the military, diplomatic and political spheres and in games. According to the Aurélio dictionary (2002), in the military sense, strategy means:

> a science which, with a view to war, aims to create, develop and make appropriate use of the means of political, economic, psychological and military coercion available to political power so that the objectives set are achieved. Another meaning is a set of means and plans to achieve an end (AURELIO, 2002).

The latter is perhaps more appropriate for the educational context.

Regarding the definition of the terms "pedagogical strategies" or "teaching and learning strategies" together with their respective theoretical references, Beluce and Oliveira (2012) provide an important explanation:

> Teaching and learning are interdependent and complementary actions that make up the educational process. The dynamic and interactive development that takes place between these actions requires the educator and the student to use strategies. The same authors point out that in recent decades there has been an intense increase in studies on the relevance of teaching strategies (PALLOFF; PRATT, 2005; ANASTASIOU; 2005, MAZZIONI, 2009; ALMEIDA, 2003; BZUNECK, 2010) or, as some authors prefer, didactic/pedagogical strategies (SALINAS, 2004; PÉREZ *et al...*), 2006), whether in face-to-face teaching conditions or *online* (BELUCE and OLIVEIRA, 2012, p. 1).

Gil (2012) highlights the fact that although the term *strategy* is used by a range of authors as techniques, teaching methods, methods, teaching activities, they are often used without much concern for determining exactly what they mean. For the purposes of this research, the term "teaching-learning strategies" in its broadest sense will include teaching methods, techniques, means and procedures. This definition was coined after analyzing the considerations of some authors on the subject.

In this sense, teaching strategies are methods or techniques developed to be used as a means of leveraging teaching and learning. For Petrucci and Batiston (2006), the word strategy has historically been linked to military art in the planning of actions to be carried out in wars, and is currently widely used in the business environment. However, the authors admit that:

> [...] the word 'strategy' is closely linked to teaching. Teaching requires artistry on the part of the teacher, who needs to involve the student and make him or her fall in love with knowledge. The teacher needs to promote curiosity, security and creativity so that the main educational objective, the student's learning, is achieved (PETRUCCI and BATISTON, 2006, p. 263).

Thus, the term "teaching strategies" refers to the means used by teachers to articulate the teaching process, according to each activity and the expected results. Anastasiou and Alves (2004) warn that "strategies aim to achieve objectives, so you have to be clear about where you want to go with the teaching process at that moment" (ANASTASIOU and ALVES, 2004).

Still in relation to strategies and their objectives, Saviani (1980) articulates the expressions "objectives" and "means" very well when he points out that:

> [...] if the objectives translate the "for what" of the action, the means translate the "with what". Both refer to the same existential condition of man. [...] In other words, if I define *this* objective, I must use *this* means or *these means'; on* the other hand, if I use *that* means, I will reach *that* objective. Therefore, there's no point in defining objectives correctly if we use means that don't lead to them. On the other hand, without defining objectives, it will be impossible to choose the right means (SAVIANI, 1980, P. 64).

For Masetto (2012), strategy in the educational field refers to:

> the set of all the means and resources that the teacher can use to facilitate student learning. Teaching-learning strategies are the art of deciding on the set of arrangements that will help the learner achieve their educational goals, from organizing the classroom space, whether in person or virtually, to preparing the material to be used, such as audiovisual resources, technical visits, the internet and other individual activities (MASETTO, 2012, p. 99).

According to Clark and Biddle (1993) the term strategy has been used in a broad sense as being integrated sequences of procedures, actions, activities or steps chosen with a clear purpose. Hyman (1987) defines strategy as a carefully prepared plan involving a sequence of steps designed to achieve a particular goal.

According to Anastasiou and Alves (2004), Clark and Biddle (1993) and Masetto (2012), teaching-learning strategies are defined as a set of actions by the teacher or student aimed at encouraging the development of certain learning competences. It is a plan of action to lead teaching towards fixed objectives, using means.

Regarding teaching skills and pedagogical strategies, Sartori and Roesler (2005) say that each of these devices requires different mediating skills and provides different pedagogical strategies, which require real-time or deferred participation, enabling expression, intervention and collaboration for the collective construction of knowledge.

Regarding the two mediations (human and pedagogical), Sartori and Roesler (2005) comment that:

> Distance Education is characterized by being a process made up of two mediations: human mediation and technological mediation, which are intertwined. The first through the tutoring system, the second through the communication system which is at the service of the first to make pedagogical mediation possible. Pedagogical mediation, resulting from the planned conception between these two mediations, is enhanced by digital convergence, which provides access and portability through increasingly integrated, fast and powerful synchronous and asynchronous communication devices (SARTORI, 2005, p. 122).

In this context of mediation, the use of pedagogical teaching-learning strategies is fundamental. Badia and Monereo (2005) state that the teaching-learning strategy:

> It is the conscious and intentional making of decisions, adapted to the context of the action, of the activity with the aim of achieving a learning objective (when the objective is to teach) can be called teaching strategies (BADIA and MONEREO, 2010, p. 363).

In order to develop teaching and learning processes, Araùjo and Marquesi (2008) reaffirm that the teacher must make use of the various existing strategies. In classroom teaching, each strategy should enable the teacher to achieve the intended learning objectives related to a given content. "Some strategies used in classroom teaching are: text study, problem solving, research, seminars, debates, directed studies, case studies.

Similarly, these strategies can be used in distance learning, as long as they are appropriate to this context and not simply transposed. Transposition should be done carefully, taking into account the resources available in the virtual environment or the use of appropriate digital resources for the strategy in question.

Vieira and Vieira (2005) classify teaching-learning strategies. Based on the principle of reality, the classification has three categories: a) real-life situations; b) simulations of reality; and abstractions of reality. The categories of *real-life* situations include the strategies of bibliographic research, fieldwork, graphic structurers, graphic organizers, networks, diagrams, flowcharts and questioning. In the category of simulations of reality, we highlight the strategies of simulation, dramatization, discussion groups, games, case studies, debates, group work, seminars, symposia, colloquia, projects, workshop, laboratory, poster, among others. The category of abstractions from real life includes exposition, readings, writing, discourse, exegesis, reading-demonstration, exposition-demonstration, programmed teaching, among others.

According to Masetto (2010), conventional strategies are those already used by teachers in face-to-face teaching, such as: group work, lectures, group drawings, brain storming, role plays, business games, research, projects, among others. In relation to technologies, the author also points out that they contribute to the development of both face-to-face and virtual education (distance learning) in the teaching-learning process. Masetto (2010) reinforces these aspects when he compares self-learning and inter-learning:

> Everything we have said about the conceptualization of the teaching-learning process in the face-to-face educational situation, we continue to assume in the distance situation. They should be used to promote *self-learning*, encourage ongoing training, research into basic information and new information, debate, discussion, dialogue, recording documents, preparing work, constructing personal reflections, writing articles and texts. Interlearning, on the other hand, is the product of the interrelationship between people. From this angle, information technology and telematics open up the possibility of debates, discussions and presentations of ideas in different times and places, with different experiences, cultures, values and customs. What a wealth of exchange!" (MASETTO, 2010, p. 154).

Anastasiou and Alves (2004), plus the recommendations of Marion and Marion (2006) and Petrucci and Batiston (2006) define teaching strategies according to the forms used: a) *discussion lists* are those that give a group of people the opportunity to debate a topic on which they are experts or have previously studied at a distance, or want to go into it more deeply by electronic means b) *problem-solving*, which is the confrontation of a new situation, requiring reflective, critical and creative thinking based on the data expressed in the description of the problem, demanding the application of principles, laws that may or may not be expressed in mathematical formulas.

According to Sigalés (2008), teachers tend to use digital resources according to their conceptions, pedagogical thinking and their vision of the teaching and learning process. Although there is a clear effort to train teachers, many of them do not use technological resources combined with didactic-pedagogical strategies in an *online* environment; or have difficulties in doing so.

Silva (2012) highlights some tools (used properly as consistent pedagogical strategies) of *online* education *such as*: *chat, forum, blog, Youtube, TeacherTube, Second Life, online* games, *Wikis* will be beneficial. Others are also highlighted by Mill (2012) when he says that among the various technologies used by teachers in distance education, the following stand out: animation, *chat*, cell phones, videoconferencing, forums, *Facebook,* discussion lists, portfolios, radio, simulators, *teleconferencing* and *webconferencing, webcam, webcast*, among others.

Silva (2010), Masetto (2012) and Dodge (2013) point out that a major driver of learning in *Moodle* are the activities, which enable interactivity in the teaching-learning process. Here are some key activities according to these authors in Table 4:

Chart 4 - *Moodle* VLE strategies and their descriptions

Strategy Name	Description
Chat	Also known as a "chat room". It enables synchronous, simultaneous contact in real time. It is an excellent channel for exchanging information, sharing ideas, clarifying doubts, etc. The purpose of the *chat* and its theme need to be well defined so that everyone can express themselves freely.
Forum	Also known as a discussion list. Used for discussions and knowledge exchange. It enables valuable asynchronous contact, as well as follow-up via *email* and file attachments. You can organize a single group to discuss or simultaneously divide the subject into several topics and form a discussion group on each one.
Quizzes	Also known in the Brazilian context as tests, *online* tests. They can be prepared in categories, forming a database in various formats such as: true or false, multiple choice, short answer, numerical answer, among others. The advantage is that the system itself corrects them, generating automatic *feedback* with simple and intuitive configuration.
Wikis	It allows the collaborative construction of documents, texts and bibliographies from the browser itself, either *online* or via a web browser. It is increasingly used in virtual courses due to its ease of configuration, the possibility of updating and, above all, its interactive potential.
Webquests (WQ)	It is a guided investigation in which some or all of the information with which learners interact comes from Internet resources, optionally. There are two types of WQ: short and long. Short WQs are those designed to be carried out in up to three lessons, while long WQs are planned for a week to a month. More than the number of lessons, this categorization is based on the skills involved. In the short WQs, the students will have had contact with a certain amount of information and, at the end of the WQ, they will have given the information some meaning. In long WQs, students will have analyzed a large amount of information, transforming this data into a product that other people can use (Dodge, 2013).

Source: prepared by the author

Studies on the use of *chat and* its advantages and disadvantages in the teaching-learning process are analyzed by researchers such as Smith *et al.* (1999); Oeiras and Rocha (2000); Pimentel and Sampaio (2001) and Hillery (1999). These authors also highlight the difficulty of using *chat* to support discussions because there are several participants sending messages simultaneously. This ends up generating several threads of conversation, all "mixed up", which can make it difficult to identify who is talking to whom and about what. This situation can be very confusing for first-time users and frustrating for users who are experienced in using this tool in other contexts, as it requires a cognitive effort on the part of the user to mentally make the cohesive links between all the messages exchanged. There are mechanisms that discuss solutions to minimize the difficulties pointed out by these authors.

Virtual Learning Environments (VLE), according to Costa and Oliveira (2004), are computer systems designed to support activities mediated by information and communication technologies. These systems make it possible to integrate multiple media, languages and resources, present information in an organized way, develop interactions between people and objects of knowledge, develop and socialize productions, with a view to achieving certain objectives. It is a virtual space that provides tools designed to allow access to a course and/or subject for distance learning and also interaction between the community involved in the teaching-learning process (students, teachers, monitors and technological support).

If face-to-face environments advocate the use of strategies and techniques that enable students to find their own meaning for the knowledge they are building with the teacher and other colleagues, Virtual Learning Environments are no different. Masetto (2012) uses the term techniques to encourage *online* learning *when* referring to teleconferencing, *chat*, discussion lists, e-mail *and PowerPoint.*

As for the Virtual Learning Environment AVA - *Moodle* (*Modular Object Oriented Dynamic), there* some fundamental strategies aimed at enhancing the entire teaching and learning process.

According to Dougiamas and Taylor (2002) and Moodle (2013), the development of *Moodle* was based on a social constructionist theory supported by four main axes: a) constructivism, a concept based on the perspectives of Piaget and Papert, according to which individuals actively construct their knowledge; b) constructionism, supported by the idea that the individual learns effectively when they construct something for others to experience; c) social constructivism, a concept that applies the above to a group that creates knowledge oriented towards others based on more complex knowledge, somehow building a culture of shared artifacts with meanings that are also shared.

Moodle integrates many of the features expected of an *e-learning* platform, including: a) configurable discussion forums, albeit to a limited extent; b) content management, allowing direct editing of documents in text and *HTML* (*HyperText Markup Language*) format; c) creation of questionnaires with the option of various types of response; d) *Chat* system configurable history log; e) *Blog* system; f) *Wiki* editor; user task management system, among others (MOODLE, 2013).

Alves and Barros (2009) point out that, just as in face-to-face teaching, teaching strategies play a key role in the pedagogical practice developed in virtual learning environments. However, the author emphasizes that these strategies, provided for in the teaching plan, must take into account the specificities of these virtual environments, considering variables such as time and space management.

Kenski (2005) understands the importance of teaching strategies that ensure proper management of the use of the different media that make up these virtual environments, with the aim of meeting the specific needs of *online* education. To this end, the author recommends planning that includes the forms of interaction and communication envisaged between users of the environment, the decision-making parameters for selecting the media and types of media support to be used, including the communication tools available in VLEs such as *chats*, discussion forums and hypertexts.

Beluce and Oliveira (2012) point out that research has also shown the relevance of strategies in and for the development of learning. Learning strategies deal with the sequence of behaviors, actions and procedures that students carry out in order to achieve a specific academic goal or perform a certain task. In this vein, Monereo (1990) points out that it is important for students to identify and implement the appropriate procedures for carrying out their study activities, as well as to develop an understanding of when and to what intensity or extent these strategies are considered significant in the construction of their learning.

In view of the authors researched, it is possible to highlight some fundamental points regarding teaching, teaching strategies and Virtual Learning Environments. For Candau (1984), Libâneo (1994), Beluce and Oliveira (2012) it is essential to understand the bilateral nature of the teaching-learning process. This means that the practice of teaching and learning are interdependent and complementary actions that make up the educational process.

Palloff & Pratt (2005), Anastasiou (2005), Mazzioni (2009), Salinas (2004), Petricci & Batiston (2006), among others, have highlighted the role of teaching strategies as the means used by teachers to articulate the teaching-learning process according to the desired activities and results with well-defined objectives.

Silva (2012), Masetto (2012), Mattar (2012), Alves and Barros (2009), among others, have addressed strategies in the context of VLEs - Virtual Learning Environments, emphasizing that they are important for pedagogical practice in accordance with good planning, taking into account the specificities of the course and the virtual environment used and the teacher's training.

5. RESEARCH INTO THE PEDAGOGICAL STRATEGIES OF E-TEC COURSES

According to Mazzotti (2001), the methodological procedures should include the indication and justification of the paradigm guiding the study, the stages in the development of the research, the description of the context, the process of selecting the participants, the procedures and instruments for data collection and analysis, the resources used to maximize the reliability of the results and the timetable. Determining the methodology of scientific research means choosing the path that will be taken to achieve the objectives proposed in the investigation. The path you choose is not entirely predictable; it can be re-evaluated along the way.

In this sense, it is essential to highlight how the research was prepared and designed. For Bourdieu *et al* (2004), research can be understood as a permanent process of construction and reconstruction, obeying the well-adjusted mechanisms of a methodological program in order to carry out fruitful work. In this way, we hope to discover ways of doing science, or of validating science that has already been done.

This capital aims to present the methodological paths of the research carried out at e-Tec CEFET-MG on the use of pedagogical strategies in the three technical distance learning courses (Internet Computing, Electronics and Environment).

5.1. Methodology

By methodology we mean the organized and planned way, following well-defined criteria, to achieve a goal, reach a solution or solve a problem. It's not enough to say where you went to do your work; you have to say why you chose that path, justify the methods and techniques you used. Methodological information must be accompanied by methods, techniques and explanations of why they were chosen (MICHEL, 2009, p. 134).

According to Michel (2009), research is classified as medium when its purpose is to identify and show how the researcher acts. This study began with an exploratory bibliographical investigation. Field research was then carried out at a higher and technical education institution in Belo Horizonte, Minas Gerais, in its distance learning department, where data was collected from teachers who taught in the distance learning modality.

The data collection techniques used were:

a) *semi-structured interviews,* with the aim of surveying some of the characteristics of the sample of teachers surveyed;

b) *documentary research*, with the aim of extracting from the Pedagogical Political Project, through analysis of the methodological proposal, principles and guidelines in order to provide a better perception for the investigation of the teaching-learning strategies of this dissertation.

After collecting the data, exploratory research was carried out in order to obtain information from those consulted and gather perceptions of the research object.

The results were discussed on the basis of the literature studied. After discussing the data, recommendations were made as to the paths to be taken in order to better guide the pedagogical strategies of teachers working with distance education, as well as suggestions for future research.

In terms of its approach, this research can be considered qualitative. Qualitative research seeks to solve questions in which it is necessary to know the reasons for certain attitudes (MORESI, 2003). However, given that this research focuses on the field of education, it could not be limited to the collection of quantitative data, as this would not allow us to delve deeper and would not provide elements for an analysis of the existing phenomenon, considering a broader view of the data. According to Trivinos (2009), qualitative research aims to obtain generalities, predominant ideas, trends that appear more defined among the people who took part in the study.

5.2. Methodological Procedures

The methodological procedures used in the research are described below, such as how the participants were selected, the instruments used for data collection, the procedures adopted for data collection and the data processing techniques.

5.2.1 Literature review

In order to carry out the literature review, it was decided to search directly in the digital library of nationally renowned universities. A number of institutions were selected: UFMG, CEFET-MG, CEFET-PR, UERJ, USP, UNICAMP, UFPE, UFBA, UNB, UCB, UFU and UFRN, as well as portals such as CAPES and SciELO. Annals of the most important national and international congresses, symposia and seminars in the fields of Education, Distance Education, Digital Media, Information and Communication Technologies were also researched.

5.2.2 Choice of Institution

In 2009, CEFET-MG set up the Distance Learning Center (NEaD), whose main objective is to introduce new methodologies that help improve teaching and learning processes. Since its launch in 2007, the Open Technical School of Brazil (e-Tec) system has aimed to offer professional and technological education at a distance and its purpose is to broaden and democratize access to free public high school technical courses, in collaboration between the Union, states, Federal District and municipalities (NEaD e-Tec CEFET-MG, 2010).

The high school technical courses offered by CEFET-MG via distance learning have NEaD on Campus VI as their administrative hub. The partner centers are run on the basis of the principles set out in Article 206, Section I of Chapter III of the Federal Constitution, which deals with education. e-Tec currently has four face-to-face support centers: Almenara, Campo Belo, Porteirinha and Timòteo (NEaD e-Tec CEFETMG, 2010).

The research site was the Distance Learning Center (NEaD) of the Federal Technological Education Center of Minas Gerais - CEFET-MG, currently housed in the CEFET-MG School Building, located on Campus VI, a public education institution in the city of Belo Horizonte.

Contact with NEaD began with the General Coordination and the respective

Coordinators of the distance technical courses: Environment, Electronics and Internet Computing. Subsequently, contact was made with the teachers indicated by the coordinators of each course for the interviews.

5.2.3 Research participants

The research used intentional samples, defined in meetings with teachers at the CEFET-MG institution where the research took place. These samples were surveyed using semi-structured interviews with teachers of technical courses in distance learning.

In order to select research subjects, Mazzotti (2001) points out that identifying the initial participants is essential for an investigation. In this study, the participants were the teachers of the e-Tec CEFET-MG distance learning technical courses.

The object of the study was the pedagogical strategies used in the teaching-learning process in distance learning technical courses. The subjects involved were teachers from the three e-Tec CEFETMG courses: Electronics, Environment and IT for the Internet.

The research proposed in this methodology was carried out with a total of six e-Tec CEFET-MG teachers, two from each course: Electronics, Environment and Internet Computing, as shown in Table 5.

Chart 5 - Teachers taking part in the research and their roles

Teachers	Titration	Training	Distance learning experience	Date of interview

PRQF_INF01	PhD student in Computer Science	Electrical Engineering and Master's Degree in Computer Engineering	In e-Tec since 2010 (3 years)	17/06/2013
PRQF_INF02	PhD student in Education and Technology	Degree in Mathematics and Physics; Master's in Technology	In e-Tec since 2008 (5 years)	24/06/2013
PRQF_MA01	Master's degree in space organization and currently finishing a specialization in teaching in distance learning.	Graduated in Geography with an emphasis on Licenciatura	Since 2010	04/07/2013
PRQF_MA02	PhD student in Geography	Degree in Geography and Master in Geography	Since 2010	22/06/2013
PRQF_ELE01	Master in Electrical Engineering	Electronics technician; Degree in Electrical Engineering;	In e-Tec for 9 months	18/06/2013
		Degree in Electronics;		
PROF_ELE02	Master in Electrical Engineering	Degree in Electronic and Telecommunications Engineering	In e-Tec from 2012	01/09/2013

Source: Prepared by the author

5.2.4 Data collection

Data collection techniques must be chosen and applied by the researcher according to the research context. For this dissertation, the following steps were adopted for data collection:

a) The General Coordinator of e-Tec CEFETMG was contacted in order to request permission for the research, as well as the telephone and *e-mail* contacts of the respective coordinators of the Internet Computing, Environment and Electronics courses. The coordinator kindly provided his contact details for this research;

b) After that, *emails* were formally sent to all the Coordinators of the respective courses asking them to nominate two teachers to take part in the interviews. The respective Coordinators nominated the lecturers;

c) telephone and *e-mail* contacts were made with the teachers indicated by their respective coordinators to arrange interview times, which were 100% granted by the teachers of the three technical distance learning courses;

d) Finally, a request was made to the General Coordinator of e-Tec as well as to the Coordinators of the three technical courses to have access to the e-Tec Political

Pedagogical Project (PPP), which was used and was a fundamental tool for analyzing the methodological proposals set out in this document in accordance with the aim of the research. The PPP was kindly provided by the course coordinators.

5.2.5 The interviews

The interview chosen was of the semi-structured type, which aims, according to TRMNOS (2009, p.146), "to compare the responses of individuals. The semi-structured interview is one of the investigator's main means of collecting data". The same author also states that the semi-structured interview in general is one which:

> [...] which starts from certain basic questions, supported by theories and hypotheses, which are of interest to the research, and which then offer a wide field of questioning, the result of new hypotheses that emerge as the informant's answers are received (TRIVINOS, 2009, p. 146).

Semi-structured interviews were carried out with teachers from CEFET-MG's three technical courses: Environment, Electronics and Internet Computing. This data collection tool can be used to collect descriptive data in the language of the subject themselves, allowing the researcher to intuitively develop an idea of how subjects interpret aspects of the world (BOGDAN and BIKLEN, 1994).

According to Marconi and Lakatos (2010), an interview is a meeting "between two people, in order for one of them to obtain information about a particular subject, through a conversation of a professional nature".

Seidman (2005) points out that, throughout the process, interviewees should be encouraged to delve deeper into their answers and share their perceptions. During the interview, we tried to identify the pedagogical strategies used by the teachers in the courses they teach.

In this dissertation, as a source of guidance for the interview, questions were used according to the interview script (Appendix C), taking into account the following categories of analysis:

a) *Teacher training and experience:* This category aimed to verify the teachers' training (technical, undergraduate or postgraduate); their experience in distance education, in the Virtual Learning Environment; how long they have been working in the distance modality, as well as their limitations in terms of knowledge of the strategies and their experience in planning, activities, preparation and selection of the content taught.

b) *Teaching-Learning in Distance Education: A* category that sought to verify teachers' perceptions of student learning; their self-discipline, self-organization; the degree to which theory and practice are linked for better learning. The levels of teacher interaction with the e-Tec Political Pedagogical Project and what pedagogical resources and strategies they have used in technical courses in accordance with the PPP's pedagogical proposal were also considered in this category.

c) *Pedagogical Strategies in Distance Learning:* Category responsible for verifying the use of the various pedagogical strategies available for teachers to use, including: *chats*, forums, learning objects, simulators, external materials, interactive activities, *webquests*, animations, *Wiki* (collaborative texts), among others. In addition, we investigated the criteria that teachers consider fundamental when choosing the tools and strategies that, in fact, provide the best results in the teaching-learning process.

The interviews took place through meetings with the teachers. The meetings with the audio recordings took place between May and August 2013, totaling six recorded and transcribed interviews.

The analysis of the interviews was divided as follows: a) firstly, the two teachers (one at a time) from the distance learning Internet Computing course were interviewed; b) at another point, the two teachers (one at a time) from the distance learning Environment course were interviewed; c) lastly, the two teachers (one at a time) from the e-Tec Electronics course. After transcription, the audio interviews were analyzed based on the categories proposed in the interview script (Appendix C).

5.2.6 Data Analysis

Mazzotti (2001) aptly highlights the aspects and dimensions of data analysis when he states that:

> Qualitative research typically generates a huge volume of data that needs to be organized and understood, and this happens through a continuous process in which we try to identify dimensions, categories, trends, patterns, relationships, unveiling their significance. This is a complex, non-linear process that involves reducing, organizing and interpreting the data, which begins in the exploratory phase and continues throughout the investigation (MAZZOTTI, 2001, p. 170).

Barros and Lehfeld (2009) point out that the success of the analysis will undoubtedly depend on the researcher himself; on the level of his knowledge, his imagination, his common sense and his theoretical-practical background and ability to argue. For the process of classification and categorization, the author also points out that the process of classification is nothing more than the division of all the data based on the establishment of a set of categories.

According to Mazzotti (2001), in order to define the units of analysis, the researcher needs to decide what interests him in a group, in different subgroups, in a community or in certain individuals. When it comes to case studies, establishing the unit of analysis corresponds to defining the case.

Qualitative research usually generates a huge volume of data that needs to be organized and understood. This is done through a continuous process in which we try to identify dimensions, categories, trends, patterns and relationships, unraveling their meaning. This is a complex, non-linear process that involves reducing, organizing and interpreting the data. This process begins in the exploratory phase and continues throughout the investigation, as stated by Gewandsznajder and Mazzotti (2001).

According to Malheiros (2011), in order to construct categories, the following suggestions should be observed: seek intimacy with the notes and transcriptions; become familiar with the data, paying attention to time; describe the data, analyzing it with caution; separate the categories by units of meaning; look for patterns; observe the nature of the typifications and perceptions; reflect on the revelations of the interview and observed data; triangulate, i.e., check the results obtained.

In this investigation, after the interviews had been transcribed, categorizations were made, taking into account important aspects of analysis and the theoretical framework used. Michel (2009) points out that the analysis of the speeches and documents at this stage should be based on the theoretical framework and aim to answer the problem and the proposed objectives. In addition, data analysis is considered:

> The richest part of the work because it directs the discussion to the real-life environment, preparing the author for real experiences in the future, it must be faithful to the information received from the object of research, be it the company, people or documents, clearly separating what is said by the researched and what is commentary by the author (MICHEL, 2009, p.142).

5.2.6.1 The Internet Computing Course

The first interview took place in the presence of two teachers, one at a time, who were teachers/authors of the e-Tec technical course in Internet Computing. Each teacher was given the fictitious name PROF_INF, as shown in Chart 6, in order to preserve the teacher's identity.

Chart 6 - Teachers of the Internet Computing Course

Teachers	Titration	Training	Distance learning experience	Date of interview
PROF_INF01	PhD student in Computer Science	Electrical Engineering and Master's Degree in Computer Engineering	In e-Tec since 2010 (3 years)	17/06/2013

PROF_INF02	PhD student in Education and Technology	Degree in Mathematics and Physics; Master's in Technology	In e-Tec since 2008 (5 years)	24/06/2013

Source: Prepared by the author

The categories discussed by the teachers of the distance learning course in Internet Computing are presented below.

5.2.6.1.1 Teacher Training and Experience - Internet Computing

At the start of the interview, Professor PROF_INF01 was asked to tell us a little about his training and teaching experience. He is currently studying for a PhD in Computer Science at the University of Minas Gerais (UFMG). As for this teacher's experience with distance learning, it can be seen that in 2010 he was invited to take part as a teacher, but the name of the course was not the same as it is today in 2013, at that time it was called Information Technology Management. He had a short introductory training course and very good support from the tutors at the time. With the help of the tutors, he tried to create quizzes to keep the student attentive to the subject content.

PROF_INF01 has been working with distance education for approximately three years. His experience with *Moodle* 1.0 was very satisfactory. However, *Moodle* version 2.0, being more complex, has caused him certain problems. Regarding the technical difficulties with Moodle 2.0, he points out:

For example, when I opened the link I could only see one of the four centers I worked at, the others I couldn't access. I had to ask the local tutor to gather information to make the correction. But I think it's a Moodle 2.0 configuration problem. I think Moodle 2.0 has a very large configuration salad and then the configuration becomes very complex, so Moodle 2.0 requires a much more advanced course. In the second semester of 2012 I came up against it a lot (Professor PRQFJNF01).

When teacher PRQF_INF01 was asked about his participation in some kind of training, he commented that although there had been a planned training course for *Moodle* 2.0, there had been no such training, either distance or face-to-face, due to the unavailability of time for the person giving the course.

When asked about any limitations in using the *AVA-Moodle* tool for the course he teaches, he made it clear that, because the tool is so broad, he wouldn't be able to extract even 50% of the functionalities it provides. He still hasn't managed to implement some of the tools, as can be seen from his speech:

For example, the *Wiki* tool. I had a bitter regret, because that tool didn't work and then it was impossible to correct the *Wiki*, because of the *Moodle* resource; if perhaps my tutor had more knowledge of *Moodle* 2.0, we might have succeeded in this problem, but we know that deep down we are in a test laboratory, in a research project." (Professor PRQF_INF01)

When it comes to *planning the course, it is* clear from the teacher's comments that the syllabus is basically contained in the course syllabus he received. Even though he can make additions if necessary, he tries to work with modules and fortnights in the course he teaches. As he himself says:

And that's how we start our planning. There are fortnights. I start chopping up the subjects so that they are more tightly packed into the fortnights. Once they've been separated, I adapt my material to those fortnights." (Teacher PROF_INF01)

As for the preparation of learning activities, PROF_INF01 said that he tries to separate his workbook into parts, as the subject is very theoretical. For each fortnight there is a reading for the student to do. The reading is the basis for their development. Within the text there are questions that refer the student to the text itself. However, none of the questions are of an evaluative nature; they function as a self-assessment. Immediately after the student has answered these questions, the teacher PROF_INF01 presents the template for the students to check their answers. To complement the theory, he posts lots of *YouTube* links. The teacher says something important about the videos:

I refer students to some important videos on the Internet. I try to turn the non-face-to-face lesson into a face-to-face lesson. Then if he takes my whole class, it will be the same quality, or at least I hope it will be at least close to the quality of a face-to-face class. (Teacher PROF_INF01).

As for the planning of the *assessments*, the teacher made it clear that the quizzes can only be taken once, there can be no more than one attempt. The teacher also tries to value participation through points for the readings and research reflections made by the student. In general, there are points for quizzes, surveys and mid-term assessments, which have a variety of questions, open, closed and related, among others. Finally, there is an assessment with a higher score at the end of the semester. When selecting subjects to teach, Professor PROF_INF01 reports that he tries to provide subjects that are as up-to-date and practical as possible.

With regard to the teacher's perception of the possible combination of face-to-face and distance learning, he reports that:

[...] it's fun. A lot of fun. The idea of being able to learn and bring technical complements from distance learning into my conventional classroom is great. And I've been reversing the process. There's a lot to improve in distance learning, but it shows that the profile of Brazilians is one that lacks initiative and discipline. It's very common to see people dispersed in the course. They think they're going to get a basic diploma. And once they realize that they need discipline to study, they start to give up. (Teacher PROF_INF01)

PROFJNFQ2, when asked if he had ever taken part in training on the AVA platform in order to work at e-Tec CEFET-MG, commented on e-Tec's initiative to offer a distance learning course for new teachers, providing technological updates on the tools. However, he stressed that the training program was short and that the rest of the knowledge was acquired from external and complementary materials.

Among the *moodle* tools, PROF_INFQ2 showed that he had some limited knowledge of the *Wiki*. The teacher explained that the tool is indeed useful, but it is necessary to understand it in order to enhance teaching. In his speech, he made it clear that it is a difficult strategy to create and apply safely.

The teacher PROF_INFQ2 uses the ideas proposed by the Meaningful Learning theorist Ausubel as a theoretical basis for *planning* the subject, relating the current content to the previous one. According to MOREIRA and MASINI (1982), Ausubel is a representative of cognitivism who proposes a theoretical explanation of the learning process from a cognitivist perspective that recognizes the importance of affective experience. For Ausubel, learning means:

organization and integration of the material into the cognitive structure, which can be understood as the total organized content of ideas of a certain individual. This cognitive structure would then be a hierarchical structure of concepts that are abstractions from the individual's experience; these concepts would be called subsuming concepts. The individual will be able to acquire meanings through the possession of skills that make it possible for concepts to be acquired, retained and appear in the cognitive structure. There is a process of interaction whereby more relevant and inclusive concepts interact with the new material, acting as an anchor but also changing as a result of this anchoring. Thus, the central idea of Ausubel's theory is that the most important single factor influencing learning is what the learner already knows (MOREIRA and MASINI, 1982, p. 41).

Professor PROF_INFQ2 also stresses that the material he used in face-to-face teaching was neither transposable nor sufficient for student learning in distance education. In this case, he tried to make various adaptations to suit the profile of the distance learning student. Regarding the student's prior knowledge in order to plan his course, he had an unexpected experience when he visited one of the centers when he said:

[...] it's not just passing on the content from face-to-face to distance learning. The question of interaction and interdisciplinarity is much stronger in distance learning. I went to a city to do a classroom lesson and I had already prepared all my material. I took a diagnostic test there. My subject has prerequisites. And this prerequisite as a subject is difficult for both the student and the teacher to teach. In distance learning, this is even more so. I really had to abandon the material I had prepared to give to the students and I realized this when I administered a diagnostic test. There was no point in me presenting the subject. I'd make the students uncomfortable and it wouldn't give good results. Because the students needed prior knowledge, as *Ausubel* said. We need this prior meaning. Based on this diagnostic test, I detected this lack of connection and attacked this piece of the puzzle. And it was fantastic, because I didn't have time to give the course, but I was calm and the students left feeling grateful because they knew that even on their own they could connect current knowledge with previous knowledge (Professor PRQF_INF02).

When it comes *to preparing activities*, PRQFJNF02 says that because he is a technology teacher, he tries to make the most of the tools within his knowledge. He uses hypertext pages, videos and a library of learning object resources. As an activity, he argues that he would like to make more videos for activities, but due to time, this possibility is compromised. She emphasized an important aspect of the student's memory by stating that she tries to work with visual and auditory aspects, among others.

In order to *assess* the students, PRQF_INF02 tries to apply objective, closed assessments so that the student can adapt. At the end of each block, the student is assessed. At the beginning of the course, it tries to give an objective assessment, but as the course progresses, open questions are added, but always taking into account the fact that some students have serious problems writing correctly according to the cultured norm.

Due to the nature of his subject, PRQF_INF01 stressed that practice is fundamental to learning. Theory is important and should be researched, but the applicability of this theory is essential. This is clear in his speech when he says: "The student must know that the content exists, look for the content, know where to look and know how to apply what he has looked for". Q PRQF_INF02 always tries to encourage students, motivating them to seek more knowledge every day and to take paths that they may not have seen before.

In relation to the criteria for selecting the content to teach, Professor PROF_INF02 proceeds as follows, according to his speech:

Whether we like it or not, we are part of this technological world. Teachers actually want their students to use all the technology they can. I want my students to have all my knowledge, but sometimes we have to hold back this pedagogical furor, this brainstorm and stop for a moment to think. What is my student's profile? Isn't it worth taking something ready-made from the Internet and throwing it at the student? That doesn't work. You have to analyze. What would be the best pedagogical methodology to apply this video, this content. Not only that, but we have to know who our client is. Each class is different. We have very different classes. In other words, for each class, we shouldn't change the content, but the application of this content to teaching. (Professor PROFJNF02).

He also pointed out that the technical course is the seed of our country, which is why schools that are really aiming for quality education should equip their teaching staff more and more in the pedagogical field. In his opinion, in some cases, teachers should always talk to the pedagogue at the institution where they work.

With regard to the perception of the relationship between face-to-face and distance learning in teaching, PROF_INF02 summarizes this issue when he says:

> The two modalities can't live on their own. If we want to get rich, we can't think of the two modalities separately. They must be applied concomitantly in their respective percentages, which will vary according to technical education. Because we know how technical education works. It's "hands on". It's knowing how to do things. Knowing how to do distance learning is no toy. EaD material in particular is challenging. [...] The combination (face-to-face and distance learning) is the key to success! (Teacher PROF_INF02).

He also stressed that the teacher's contact with the student in the classroom is essential. And in face-to-face teaching itself, it is possible to use video lessons or other strategies that are also used in distance learning to promote quality teaching.

5.2.6.1.2 Teaching and Learning in Distance Education - Internet Informatics

As for Professor PROF_INF01's expectations regarding student learning, he emphasizes that face-to-face visits are fundamental for learning. In addition, the teacher should try to get *feedback* on student learning through the *AVA-Moodle* tool itself. The teacher believes that the profile of Brazilian students is complex and that this often leads to poor expectations from the teacher's point of view.

From PROF_INF01's point of view, the difficulties in distance learning are different from those in face-to-face courses. Face-to-face students talk a lot, whereas in distance learning there is no such thing, except for the activities. The idea is to encourage interaction between two or more students so that they can talk in person.

As for the perception of the relationship between learning theory and practice, teacher PROF_INF01 emphasizes that practice in the classroom is guaranteed. Distance practice is not guaranteed, it is optional. Even so, the teacher tries to encourage students to practice on their own according to the context of their subject. It is possible to simulate various situations in the subject in order to improve student learning.

In order to accompany the construction of the student's knowledge, teacher PROF_INF01 relates an important fact:

> I can't guarantee that I'll monitor the construction of the content. I measure as best I can. The questionnaire and the forum are tools. I check these kinds of things. I look at the students' forums and try to keep up. Originally I had 150 students and suddenly it went up to 70. That's a lot of students to keep up with. (Teacher PROF_01)

As for his *knowledge of* and level of *interaction* with the methodological proposal of the e-Tec Political Pedagogical Project, the teacher stated that it was a reference with which he had some knowledge at the beginning of his work. It was just an original guideline. The teacher's statement makes it clear that the PPP is not unknown, but it is not his focus.

Among the various resources set out in the PPP, the teacher clearly stated that he doesn't like *chat* and prefers the forum to *chat.*

Professor PROF_INF01 emphasized that he is not currently using the original idea of the methodological proposal of the Pedagogical Political Project with 12 units of study, but is instead working with eight units. He added that each unit provides: texts, quizzes, lots of links to videos and up-to-date support material.

When PROF_INF02 was asked about his expectations with regard to student learning and himself as an e-learning teacher, he highlighted that the course

none can give 100% professionalization according to their experience as a teacher. It's a certain amount of direction, instrumentalization and, finally, the student must find the rest in the field of work. Many of the resources the student will find in the workplace.

Professor PROF_INF02 believes that one of the aspects of an efficient distance learning student is discipline. If students don't plan their time in distance learning, they certainly won't succeed. If they don't have discipline, they will inevitably drop out of the course. A distance learning course depends more on the student than on the teacher. The teacher in distance learning stimulates, but the student needs to have autonomy. The teacher, on the other hand, highlighted something important about discipline when he said:

That student who participates, who is always on the AVA platform, and because of some cognitive learning problem hasn't got there yet, that student deserves all the attention of the teacher in question. It's stressful for the teacher, but they deserve it. We have to give water to those who want to drink it; those who don't want water don't want water! No matter how tasty my water is, no matter how much you encourage them to drink it by saying: water is good for you, but if they're not thirsty, how are they going to drink it? The teacher must respect the student. Sometimes they don't really want to learn (Teacher PROF_INF02).

PROF_INF02, when asked about his perception of the difficulties faced by face-to-face and distance learning students, pointed out that, first of all, every teacher should teach in both modalities. When teachers learn to teach in distance learning, they become better teachers in face-to-face teaching, and the opposite is also true. With regard to the study discipline that the student should have, Professor PROF_INF02 brings up an approach that differentiates the modalities with regard to the study discipline expressed by the student:

[...] There are two moments in the classroom, for example. The first moment is when the student's father says "it's time for you to go to school". Then he sits in the van and the student goes to school whether he wants to or not. The second moment is when the student leaves school and has activities to do at home. In this case, they may or may not be driven by their parents. The teacher leads in the classroom and the parents help lead at home. In distance learning, this changes a little. Because in distance learning it's the student themselves who has to lead their studies. In the classroom, the student has a timetable to follow. In distance learning, they decide... am I going to do it when I get home from school, at night or early in the morning? Anyway, I think that in distance learning the student's concept of study discipline has to be formed. The level of maturity of the distance learning student is something that needs to be worked on (Professor PROF_INF02).

In the same context, Professor PROF_INF02 drew a parallel between distance education and the army when he said:

The distance learning student who does well is the student who served in the army. Because there they work hard, discipline themselves and come out alive. Alive in distance learning are those who leave ready for the job market. With seriousness, discipline, maturity and autonomy. That's if they don't already come with this background. (Professor PROF_INF02)

When professor PROF_INF02 was asked about teaching and student learning between *theory and practice* in distance education, he highlighted the fact that it is essential to promote technical visits in practical subjects, equalizing learning. More theoretical subjects should also have this concern. With regard to distance learning in the e-Tec research field, the professor observes, from a pedagogical point of view, that the institution still needs to "wake up" to distance learning, due to the fact that it is a worldwide reality. Awareness of the change is fundamental. There are a number of government initiatives to bring technological "roads" to the centers. Several institutions have already received these resources from the government in a more transparent way than this one. In this case, the teacher denounces the lack of political commitment on the part of the institution to boost the teaching-learning process in distance education. This is evident in his speech when he says: "The practice exists, but we also need to take it to the people, to the centers". The government has technological "roads" with everything ready. The less privileged will have access to the practice, but you have to be proactive in this process.

In order to monitor the student's learning, PROF_INF02 reported that he makes a diagnosis to see if there are any doubts about the previous content so that he can move on to the next subject. In this context, *feedback* is important, since it is possible to check whether the student can continue in the process of building knowledge or whether the learning needs to be resumed.

As for his *knowledge of* and level of *interaction* with the methodological proposal of the e-Tec Political Pedagogical Project, PROF_INF02 said that he had helped to develop it during the construction of the course. For him, interacting with the PPP proposal has contributed to improving the quality of classes and teaching-learning processes.

5.2.6.1.3 EaD Pedagogical Strategies - Internet Informatics

Regarding the pedagogical strategies used, an important observation about video lessons was made by Professor PROF_INF01:

[...] due to the nature of the subject, the videos must be made available with great care. We have a serious problem with resources at the hubs. I'd like to do video lessons of questions, for example. I'd like the hub to connect with me live and see my lessons. Do a simulation, but I don't have the bandwidth for that at the centers. We've had problems with *Youtube.* Pointing out *links* on *Youtube* is very good, but complicated, because everything on *Youtube* is copyrighted. We thought we'd send it to the hubs and have the students access these videos via a local server. Students often spend an hour to an hour and a half watching a 15-minute video. This is very complicated for the student. In other words, the city needs to invest in infrastructure. If the city is applying for a hub, it should make this structure available. There are many very good tools, but they need bandwidth (Professor PROF_INF01).

As for the use of simulators, the teacher said that he would need more appropriate training to produce this type of strategy. According to him, e-Tec has something related to this, but due to lack of time, he wouldn't be able to edit videos and treat them with adequate precision.

With regard to *Learning Objects*, teacher PROF_INF01 said he was unaware of this type of strategy. He points out that, in this case, the teacher's role is to choose the video on *YouTube* very carefully. He believes that these videos are better than any animation. Animation would be essential to show situations behind something that people don't see clearly. Animation in this case would be important, he said.

Texts are among the strategies that teacher PROF_01 mentions as being the most effective for students to learn on their own. This is because students need to read a text more than once. Thus, when the student reads a text three times, the first reading guarantees contact with the knowledge; the second is aimed at putting the pieces together mentally and the third leads to learning the content. Secondly, the strategy that brings the most learning, in the case of the context of his subject, are the videos that he carefully makes available to the student on the AVA. The teacher himself tries to read his own texts for self-study three times, thus guiding his students in the same direction.

Professor PROF_INF01 was also asked about the criteria for choosing strategies within the VLE as a way of enhancing learning.

At this point in the interview, he declared:

First of all, I use the theoretical framework with the text. Then, if I'm going to use *Wiki,* for example, I choose to take the focus off the questionnaire. So I change the dynamic. [...] With the Forum, I try to leave one Forum open and another that relates to that fortnight or week. Usually, the students participate a lot. But the idea is that we (as teachers) bombard the student with the Forum. The criterion in this case is to promote the deepening of some subject that hasn't yet been explored in more detail. In the Forum, the criterion is to promote in-depth study of the subject. On the *Wiki,* it's to energize and motivate the student. When the student gets used to texts and videos only, I go there and use a forum or *Wiki*, so that the student doesn't fall asleep

(Professor PROF_INF01).

Professor PROFJNFQ2 stated that he has not used *learning objects* and *WebQuests* as teaching strategies in the VLE. Simulators have not been his focus and *Wikis* have been used infrequently. He pointed out that although *Wikis* are a collaborative strategy, it is necessary for the teacher to include bibliographical references of a scientific nature to guide the students, since there are sites with no scientific relevance for research and subsequent insertion into the *Wiki* by the students.

The strategy that PROF_INFQ2 highlights as a major learning factor for students in distance learning is the Forum. Because questions are shared and other people, such as the teacher himself, are able to express their opinions. It's a feedback tool. It is always being "fed" and always "feeding" knowledge. The teacher believes that this in fact motivates autonomy on the part of the student.

As for the most used strategies, Professor PROF_INFQ2 again highlights the use of hypertexts. He makes a lot of use of hypertexts, videos and *Power Point* presentations with lots of effects to make the lesson more dynamic. However, he points out that he would like to use other teaching-learning strategies.

Professor PROF_INFQ2's *criteria* for choosing strategies are based primarily on the content he is going to teach. It is the content that determines the strategy or strategies and not the other way around. In very theoretical subjects, he generally uses forums, emphasizing that the discussions are of an investigative nature, i.e. in the forums the students must seek out the information, bringing it back not in its entirety, but in order to discuss it on some fundamental point. He cites an example:

In a theory course, I can do some research on the father of HTML. Then I ask the students: Who was he? What did this guy do? Do you know him? What good did he do? You have to create movement in theory. Different from the practical part. They need to be more focused. You have to give him (the student) tips (Professor PROF_INF02).

The student's life is part of the PROFJNFQ2 teacher's interests, because in this way he is able to make the information transmitted more applicable. It is also important to know what area the student works in so that it is possible to juxtapose what the student learns with their work and life context. Knowing the profile of the students precisely because it is distance learning is essential for successful teaching to reinforce PROF_INFQ2.

5.2.6.2 The Environment Course

The second stage took place in the presence of two teachers, one at a time, who were teachers/authors of the eTec technical course in the environment. Each teacher was given the fictitious name PROF_MA, as shown in Chart 7, in order to protect their identity.

Chart 7 - Teachers of the e-Tec Environment Course

Teachers	Titration	Training	Experience EaD	Date of interview
PROF_MAQ1	Master's degree in space organization and currently finishing a specialization in teaching in distance learning.	Graduated in Geography with an emphasis on Licenciatura	Since 2Q1Q	Q4/Q7/2Q13
PROF_MAQ2	PhD student in Geography	Degree in Geography and Master in Geography	Since 2Q1Q	22/Q6/2Q13

Source: Prepared by the author

The categories discussed by the teachers of the distance technical course in Environment are presented below.

5.2.6.2.1 Teacher Training and Experience - Environment

Professor PROF_MA01 has ten years' experience in distance education. He started with e-Tec in 2010, when he was invited to set up a tutorial for a subject and be a content teacher for some other subjects in the program. "I had never had any contact with distance learning until then," he says.

Professor PROF_MA02 has worked as a classroom teacher since graduating. He started at e-Tec in 2010 as a Tutor and, in 2012, took on subjects as a teacher.

According to Professor PROF_MA01, the first semester was very difficult, despite the fact that he had been partially prepared for the tutorial, but not for teaching in distance learning. He had some guidance from the pedagogical coordinator and the technical team on how the platform would work and some basic tools. In the first year, he was well assisted by the team, but there was no specific training on how to teach distance classes. Regarding the assessment process, he points out that it was hit and miss. In the first course, there were several problems and many difficulties.

PROF_MA02 said that tutoring itself brought experience to his training and preparation as an e-Tec teacher.

As for his experience with AVA *Moodle,* Professor PROF_MA01 said that at the beginning it was proposed that he should focus on tutorial production, and for this there was training funded by the MEC in another city. There was very specific training to produce the tutorial, but not to work in *Moodle*. You can see this when he says:

They would just come in and tell me that I had to select a variety of video and audio media. Finally, in addition to texts and hypertexts, websites, where I would pass this on via e-mail to the tutor and the tutor would make the posts. I simply arrived with the Moodle *layout* for my course already pre-prepared. I simply operationalized the course in terms of content; and in technical terms, the tutor did everything herself (Professor PROF_MA01).

PROF_MA02 said that there were several training courses at the beginning. These courses were taught at a distance; however, the questions were answered in person at e-Tec.

PROF_MA01 reported on his difficulties and limited knowledge of the VLE tool at the beginning of his work:

In the beginning I had a lot of difficulty because I didn't understand all the tools, you rely on what you know. In the face-to-face model, the formal exams, that part was easy to set up, the forms were already ready, the format was already ready, all I had to do was define the number of questions and basic alternatives. So when it comes to preparing tests, we use the face-to-face model, but when it comes to teaching classes and monitoring students' learning, diversifying the use of tools, I did have difficulties. I think not just me, but the students too (Professor PROF_MA01).

Regarding *chat* specifically, PROF_MA01 also declared his major difficulties:

At first, the *chat* was compulsory, we had weekly meetings and I had to spend at least three hours in the weekly chat. Without any prior guidance. I was supposed to be available in the chat as an answer to questions. Two modules later, in the transition from 2011 to 2012, it became clear that the *chat* was not an effective tool, few people used it and when they did, they used it in the wrong way, to chat for nothing and not on topics related to the course being taught. Some used it, but if you created a thematic forum, which was the ideal, it was even a guideline from the pedagogical coordination, you didn't have the capacity to interact with 10, 15 students simultaneously: the video would drop, or you wouldn't be able to have a linear conversation. So we came to the conclusion that *chat* was not a tool that was used well. There was no satisfactory feedback until it became non-mandatory. I can even use it as long as it's requested by me or by the student themselves, either individually or in a more organized way, but I have to comply with these limitations (Professor PROF_MA01).

PROF_MA01 pointed out that, due to lack of knowledge, he doesn't use *podcasts* (audios), but he would like to use them in his next course. He said he would also like to do some video interactions. Homemade" videos, since the current physical structure doesn't have a suitable video room. The limitation that PROF_MA02 said he had with regard to tools was basically with the *Wiki.* He reported that he had used it at one point in his course, but the result was not appropriate.

To *plan the subject*, PROF_MA01 organizes his classes on a fortnightly basis, according to the topics covered. the topics covered the tutorial, against the backdrop of the tutorial provided, the support materials; complementary activities, complementary media, complementary texts, always seeking to diversify the language.

PROF_MA02's planning is based on the tutorial offered at the beginning of the course. He tries to divide the workbook into fortnights with a semester timetable and adds closer themes for the preparation of the assessments.

The *preparation of the material* depends on the subject, according to PROF_MA01:

> I'm currently teaching, linked to e-Tec, four subjects, two of which I put together myself. The other two don't come from the MEC's own database, or from other e-Tec teachers, who also put together the material. Because it depends a lot on demand, in principle there can't be two teachers teaching the same subject at the same time. So it could happen because of the organization of the entrance exams and it's not yet continuous and it could create a hole, a gap, so there could be an overlap of subjects per module (Professor PROF_MA01).

In order to design *learning activities,* PROF_MA01 reported that, at the beginning as a teacher, he often used the multiple choice questionnaire format. However, as time went by, he tried to investigate "wikis*",* despite the difficulties in assessing students in general. Individually, she had difficulties with this tool, but stressed that she intends to use it again when she has prepared herself better.

To prepare the activities, PROF_MA02 said that most of the time he prepares them based on the theoretical and reading content. Generally, he balances multiple choice. He uses open questions and true/false. *He* also uses articles and *YouTube* videos.

Regarding the *Forum* strategyPROF_MA01 pointed out that:

> In almost all subjects, I think it's one of the most important tools for distance learning. When you have access to interaction with the student, you can follow their reasoning, that of the class, you have a specific place to give *feedback* which is made available to the student and to other colleagues. So one colleague reads the other's answer and he learns, so the feedback can be both individual and collective. I think that in this sense it is a tool that is used a lot and I always use it (Teacher PROF_MA01).

PROF_MA01 made it clear that, at times, he uses the quiz as a way of bringing the student closer to the final assessment process, since it's not only interesting to create collaborative activities, but also text preparation activities. According to the teacher PROF_MA01, the test can be factual or in quiz format so as not to clash too much. So he tries to balance it with research activities, bringing the topic and the characteristics together, because the environment course allows this.

With regard to the *preparation of activities and the correction of assessments,* PROF_MA01 stated that, at first, it was very factual; then, the way he saw the distance learning student changed a lot, especially after he visited the centers. In his speech, his perception and surprise about the students' profile is clear:

> We assume, and expect, the greatest difficulty for teachers in this modality. Because we idealize a student profile. Even more so as a CEFET student. You imagine that it's a student who has time available, who has had an excellent education in primary and secondary school and who is able to develop the topics covered. On a visit to the hubs, I saw that they are not (Professor PROF_MA01).

After that, PROF_MA01 started reformulating his tests. Regardless of the type of question, the corrections were made directly in the environment. Teachers take a closed test in *Word* and then set it up in the AVA. This way, the student takes the test and gets the result immediately. For the final exam, which is face-to-face and printed, PROF_MA01 tried to combine other types of questions, creating easy, intermediate and complex ones. As for the scoring, PROF_MA01 believes there is a balance between the assessments because one is only multiple choice and the other varies the type of questions, such as open descriptive, analytical or argumentative, at which point the student tries to argue about the topic in question.

In order to *select* the teaching content, PROF_MA01 pointed out that he first follows the syllabus, associating it with the skills and competencies that the technical course student needs. He also highlighted the preparation of interdisciplinary reports which the student has to hand in at the end of the semester. This is an interdisciplinary and transdisciplinary report aimed at learning for the job market.

For PROF_MA01, face-to-face meetings are essential for effective learning. It is necessary to get to know the reality, the context and the students in order to then adapt the teaching methodology to the educational environment.

When PROF_MA02 was asked about the possible combination of face-to-face and distance learning, he reported something important:

> [...] I don't think combining would be a good word. I think the modalities are different: distance and face-to-face. The face-to-face methodology won't work in a distance learning course. [...] Technical education already says a lot. The name says it: technical. Something more practical. When they created the electronics course, it was a mess. I don't know how it is today. The distance learning Environment course, I think in general, is quite complicated. In face-to-face teaching, students have difficulties. At a distance, the difficulty is double and answers minus two. Not just because of the laboratories and materials, but I think the presence of the teacher is lacking. I can feel it in the students. And because it's at a distance, I can't help any more. In general, the lack of a teacher is a complicating factor (Professor PROF_MA02).

5.2.6.2.2 Teaching and Learning in Distance Education - Environment

With regard to student learning in distance learning, PROF_MA01 said that the main advantage was flexibility. Students from the outskirts, who don't have access to technical education, would now be able to take a course. However, this modality requires other attitudes from both the student and the teacher. From the student, preparation, and from the teacher, planning, as PROF_MA01 says:

In distance learning, there's no room for immediacy. In face-to-face classes, I walk into the classroom and suddenly there's an insight. It's not like that in distance learning. The teacher needs to structure beforehand. In this model, the advantage that the student has is that he has training that is compatible with face-to-face teaching, and he does it at his own time and at his own pace. I think this is a great attraction, but I also think that there is still a long way to go. Especially in Brazil, in terms of technology and organization. We're not used to the profile of an autonomous student. This is something we have to break through if distance learning is to move forward.

PROF_MA01 pointed out that the advantage of distance learning is that teachers are able to work from home, without the natural stress of a classroom. Another advantage for teachers is that they can make use of more emerging technologies. As a disadvantage, PROF_MA01 highlighted the issue of salaries and the scrapping of distance education. Distance learning teachers don't have the CLT guarantees that face-to-face teachers currently have. When comparing face-to-face teaching and distance learning, PROF_MA01 pointed out that, from a professional point of view, he would make all his time available for this modality, but thinking about the financial side and making a living, this is not yet possible, and it is also necessary to keep up with face-to-face teaching.

For PROF_MA02, distance learning in the current e-Tec format, in the specific course he teaches, has not met the student's expectations. Especially with regard to the doubts that students have in general. The higher the degree, the more experienced the student seems to be, the better they learn. For younger students, however, distance learning doesn't really meet their needs.

Regarding his expectations as a teacher in distance education, PROF_MA02 reveals something essential in his speech:

I'm a bit frustrated. It's sad for the teacher to teach and the student not learn. At the same time, I don't think it's the teacher's fault, but the student's lack of interest. Perhaps if it were in classroom teaching, this would be easier. This distance makes learning difficult. I think I owe it to myself to teach, but because of the difficulty of this distance relationship (Teacher PROF_MA02)

Regarding PROF_MA01's perception of student *discipline and autonomy* in distance learning, he stressed that this issue involves both the student and the teacher themselves.

It's important to have a disciplined student, but also a teacher who motivates that student. It's important to have a good platform, good content, but there must also be a lot of teacher involvement. Partnership on both sides: both the student and the teacher. PROF_MA01 reinforced that it is essential for the teacher to stimulate the student on the platform. This is evident in his speech when he says:

Teachers can't just abandon their students. Let the tools speak for themselves, because they won't do it. We have to constantly encourage students to get to know the platform; make them want to spend more time on it. Both the student's profile, so that they are autonomous, but also the teacher's, the team's, the faculty's, teaching-learning strategies are required to make them feel motivated and attracted to stay on the course (Professor PROF_MA01).

For PROF_MA02, a student's biggest difficulty is the issue of planning. Students find it difficult to dedicate time. It's rare for a student to dedicate at least four hours to learning on e-Tec. There are considerable differences between the difficulties faced by face-to-face students and distance learners. PROF_MA02 points out that if the student is a disciplined person, they will have few difficulties with discipline and autonomy in distance learning. In his speech, we can see PROF_MA02's concern about the student's priorities in education and distance learning:

My perception is that students always put education second. If the student's first plan is education, then everything will be fine. In general, if the guy is slack with other priorities in his life, then he'll be the same with education and distance learning. People who are dedicated to their studies are generally dedicated to other things in their own lives. You rarely see a person who is sloppy in life and dedicated to their studies (Professor PROF_MA02).

When asked about his perception of the relationship between *theory and* student *practice* in distance learning, PROF_MA01 reported that some students manage to associate theory with practice. At times, according to his experience, PROF_MA01 has noticed that there is a certain unevenness on the part of the students. There are students who have enormous practical experience but little theoretical knowledge, and vice versa. They have different professional goals, different contexts and this becomes a factor that hinders learning, which needs to be investigated.

For PROF_MA02, the proposal to have fieldwork (practical) in the theory is good, but this hasn't been happening. More than 90% of the content is theoretical and not practical. For a technical course that requires something tangible in the distance learning mode, learning is really lacking.

In order to *monitor* the construction of students' knowledge, PROF_MA01 initially observes the forum activities; how the student argues, contextualizes, absorbs and dialogues with the content. He also tries to observe the assessments and how the student argues his doubts in the specific forum. In view of this, it is not possible to make a good diagnosis of students with little participation. PROF_MA01 also stressed the importance of interaction between the content teacher and the tutor. This is clear from his speech:

Because sometimes the tutor knows a lot about the operational side, but doesn't know the content. Then the teacher steps in. There isn't much dialogue between the teacher and the tutor at the pole. I think that would solve a lot of problems. To meet, to talk. Because the tutor has direct contact with the student. The teacher does, but at a bit of a distance. The tutor could come to me and say: look, the students are having difficulties with this approach. Then, as a teacher, I'll adjust it. Generally, we teachers only realize this late. It could be done earlier (Teacher PROF_MA01).

The way in which PROF_MA02 has accompanied the construction of knowledge is basically through activities in the VLE environment. He reported that he had tried to use *chat,* but had been unsuccessful with this strategy due to problems of compatible timetables between the students and himself.

With regard to knowledge of and interaction with the Pedagogical Policy Project, PROF_MA01 reported that it was during this interview that he had contact with and knowledge of it. From 2010 until the time of this dissertation, PROF_MA01 had had no contact with the e-Tec PPP. He stressed that he was aware of the syllabus for the subject he teaches, but only in a very fragmented way. He only said verbally that he had had contact with the PPP document through the coordinators.

Of the strategies in the e-Tec PPP's methodological proposal, PROF_MA01 said that he hasn't used video lessons because he doesn't have a structure within eTec for this, but he would very much like to. He has only researched and used them initially.

For PROF_MA02, the Pedagogical Policy Project is unknown. He pointed out that he doesn't know if it's because he himself as a teacher hasn't sought it out at meetings or if it hasn't actually been presented to him. Specifically, teacher PROF_MA02 stated that he had not had access to this document at the time of the interview.

From the PPP's methodological proposal, PROF_MA02 has tried to use video lessons and support texts posted on the AVA in his classes. Although he hasn't used games, he is interested in using them. He also pointed out that complementary activities such as fieldwork would be one of his interests.

5.2.6.2.3 Pedagogical Strategies in Distance Learning - Environment

The strategies that PROF_MA01 has used the most are forums, questionnaires and external materials on *YouTube.* At the time of the interview, the teacher was unaware of *Learning*

Objects and *WebQuests*. Among the strategies mentioned by PROF_MA01, the ones that stood out as providing greater autonomous learning on the part of the student were the collaborative tools: *forum*, *Wiki* and video lessons. When it comes to assessing collaboration, PROF_MA01 believes that the *Wiki* is more laborious due to its subjectivity, while the forum is easier to diagnose collaboration.

PROF_MA02 has mainly used the forum strategy. *Chat*, he said, is not a good strategy to use. He prefers the student to post the question on the VLE, because the student asks the question and the teacher clearly and calmly prepares a good answer. Learning Objects are totally unknown to PROF_MA02. Simulations and animations have never been used by the teacher. *WebQuests* have already been used in certain lessons. He used the *wiki* at a certain point in the course, but reported that he couldn't get all the students to attend the VLE at certain times. As a result, he had to redo the whole process, which caused him considerable stress.

The same teacher PROF_MA02 pointed out that *WebQuest* is the strategy that most promotes student learning. The forum would be in second place, followed by the *Wiki*. PROF_MA02 said that the same material that the student has in hand is uploaded to the VLE, in PDF format, every two weeks. He basically uses video lessons, articles on the Internet and forums in his classes.

Regarding the criteria for *choosing* the strategies used in class, PROF_MA01 initially stated that, in the first place, his own limited knowledge of the tools becomes a criterion for choice. Generally, he tries to choose a different questionnaire format with open questions, *cartoons* to interpret, as well as trying to associate figures and reports so that the student tries to develop a critique of the content presented.

PROF_MA02 reported that he uses the forum for theoretical content. However, when the content is simpler, he uses multiple choice questions. For the student to reflect a little more, he tries to use open questions.

Among PROF_MA01's reflections, the ones that stood out were: a) teachers lack a lot of training and education. In a humorous way, he declared that one of his dreams would be "not to need a tutor". Although the tutor is fundamental to the process. b) showing teachers the pedagogical possibilities of the tools available for effective use by teachers in their teaching context.

As suggestions for improving the teaching-learning process in eTec, PROF_MA02 stressed the need for greater proximity with the course coordinator and teachers. Meetings should be more frequent. With regard to student difficulties, PROF_MA02 clearly stated that:

> Some of the difficulties that the students themselves have been facing are exactly a reflection of the difficulties that the teachers themselves have been facing. And perhaps the coordinators themselves in other instances, I can't say for sure. Just as the teachers have to dedicate themselves to their lessons; 50 minutes a week. I think there should be more proximity between the coordinators and the teachers. At the start of e-Tec it was good. There was one person who did some pedagogical work, but I don't know what happened, everyone disappeared (laughs). This is all well "distance" (Professor PROF_MA02).

He also said that it was essential for teachers to evaluate their own lessons with the participation of the coordination team. He reported that he did not receive any feedback on his own lessons from the coordination team. In order to improve as a whole, PROF_MA02 proposes that the teachers themselves should be held to a higher standard in terms of their work, charged more rigorously and assisted more closely with *emails,* messages, etc. by the course coordinators. The pedagogical team should suggest more the use of strategies by teachers with tips and guidance. The two meetings per semester generally revolve around specific problems that are necessary; however, more "pedagogical" meetings would be needed to help teachers in their daily work with strategies.

At the end of the interview, PROF_MA02 was asked about his perception of distance learning. The teacher replied:

> I'm still in the process of believing in distance learning (laughs). I have a particular vision from what I've seen from the students. I think that, in principle, distance learning can work, or has been working at the higher levels. At the technical level, there's still a long way to go in terms of students. At undergraduate and postgraduate level, it's easier. Distance learning has to keep pace with the student's maturity. Distance learning is actually a response, a reflection of the student's maturity. The idea of bringing technical education to people who have little access to it is a good one, but from what I've seen, it hasn't worked! (Professor PROF_MA02).

5.2.6.3 The Electronics Course

The third moment took place in the presence of two teachers, one at a time, who were teachers/authors of the e-Tec Electronics technical course. Each teacher was given the fictitious name PROF_ELE, as shown in Chart 8, in order to preserve the teacher's identity.

Chart 8 - Teachers of the e-Tec Environment Course

Teachers	Titration	Training	Distance learning experience	Date of interview
PROF_ELE01	Master in Electrical	Electronics Technician; Degree in Electrical	In e-Tec for 9 months	18/06/2013

	Engineering	Engineering; Degree in Electronics;		
PROF_ELE02	Master's in Engineering	Degree in Engineering Electronics and Telecommunications	In e-Tec from 2012	01/09/2013

Source: Prepared by the author

The categories discussed by the teachers of the distance learning course in Electronics are presented below.

5.2.6.3.1 Teacher Training and Experience - Electronics

When talking about his training and experience, PROF_ELE01 reported that he has a degree in Electrical Engineering, as well as a Full Degree in Electronics with a Master's Degree in Electrical Engineering from UFMG. His experience in distance education is approximately nine months.

Professor PROF_ELE02 has a technical degree in Electronics from the Technical College of the Federal University of Minas Gerais - COLTEC/UFMG. His degree is in Electronic and Telecommunications Engineering from the Pontifical Catholic University of Minas Gerais - PUC-MG, with a Specialization in Maintenance Engineering from the IEC - Institute of Continuing Education and a Specialization in Telecommunications from the same IEC with a Master's Degree in Electrical Engineering in the line of research: Artificial Intelligence - Evolutionary Systems.

With regard to training in distance learning, PROF_ELE02 pointed out that he had no training to work in this modality, just training or guidance on how to use the platforms. The first contact he had with distance learning was at the Brazilian telecommunications company Oi, in 2006, when he took part as a student in training courses using this modality. Later, in 2008 and 2009, he became an instructor for two distance learning courses at the company. In these courses, he explained the processes of billing and tariffing telephone bills, an area in which he was a specialist. He also took part in three distance learning courses through *Coursera*, at *CalTech* and *MIT*, in the areas of artificial intelligence and robotics.

In the first semester of 2012, he was invited by the e-Tec coordinator to take part in the selection process for a teacher of the Embedded Electronics subject, as he was a substitute teacher for the electronics technical course and the post-graduate course in Electronic Systems and Automation, working exactly with the syllabus covered in the subject.

Subsequently, with the good evaluation of the results of the course, he also took part in the selection process for teacher of the course End of Course Project at EaD-CEFET-MG.

Teacher PROF_ELE01 says that e-Tec provided a training course on the AVA - *Moodle.* However, he didn't take part, as he was only going to be a facilitator. As for using the AVA, he says:

I don't use the VLE very often. I use the VLE basically for *chat* and to enter an activity, but even then I ask the tutor to take a look to see if it's up to the standard he's used to. My limitations exist. I think I've settled on what I think is necessary at the moment, which is the *chat* I always do. I know there are other activities, but I don't know about them" (PROF_ELE01).

PROF_ELE02 reported that he took part in a basic orientation course with his tutor as soon as he started on the distance learning program. He researched and discovered some resources on his own, asking the tutors when he needed a resource he didn't know about. She pointed out that unfortunately she was unable to take part in the one-day course in July 2012, as the deadline was short and she had already taken on other commitments.

PROF_ELE02 made an important criticism and proposed suggestions with regard to the *VLE-Moodle chat* strategies adopted so far*:*

My main criticism is the unavailability of the excellent *Open Meetings* tool that allows virtual classes, where the student can interact with the teacher in a much more efficient way than typed *chat.* It is a tool for virtual classes, which brings the student closer to the teacher. I tested the tool for a few weeks and the student approval rate was so high that I started using an external tool (*TeamViewer*) to interact with the students on a weekly basis, replacing the *chat,* which proved to be inefficient and unproductive. In fact, during the platform's long hiatus in the second semester of 2013, we continued the course and working directly with the students using this tool. Another consideration would be to standardize the organization of the fortnight on the platform for all subjects, as each teacher adopts their own way, making it difficult for the student. The Notes section also sometimes has problems and is very unproductive (Professor PROF_ELE02).

PROF_ELE02's awareness of the need for strategies other than the traditional *Chat is* perceptible. This corroborates the allusive words of Perrenoud et al. (2001, p. 80) when he points out that "the teacher's practical knowledge must be contextualized, embodied and finalized, transforming it into knowledge adapted to the situation".

Teachers must interpret their experiences on the basis of both theoretical and practical knowledge of their work.

When asked how he plans his subject, PROF_ELE01 reported that his planning is basically based on the "Course Program" and eight-week timetable he received. He plans using a linear division in terms of difficulty. In his speech he explains what he means by linear:

For students to know about analog electronics, they need to know about electrical circuits. I see the need for this. In the first fortnight I cover the basics of electrical circuits. In the second week I moved on to semiconductor components. In the third, I link more specific components. In the fourth, the application of this specific circuit and so on. I say linear in importance. The subsequent thing builds on the previous one. It's always like that. Not in terms of difficulty. But as the subject grows, more knowledge is added. What you have in week five adds knowledge from weeks four, three, two, one (PROF_ELE01).

Basically, the teacher PROF_ELE01 uses the e-Tec "Course Program" as well as handouts from another Federal University with additional notes made by himself to plan.

PROF_ELE02 said that he plans his course by first discussing the syllabus of the course with the course coordinator, and the availability of practical and face-to-face classes. He then tries to put together the whole course plan with the content to be covered each fortnight, the timetable, the course content handout, essential supplementary material, assessment criteria and the standard for posting on the platform. He has a meeting with the tutor, when he tries to exchange ideas and explain how he wants to conduct the course. "We discuss and define the roles and responsibilities of each person and how we will interact with the students to answer their questions," he says. A deadline is set for the fortnight's posting. As soon as the fortnight begins, the planning for the following fortnight is revised, with content being added or deleted.

The teacher's management domain is fundamental to making the teaching-learning process effective. According to Behar (2013), there are domains that teachers must pay attention to in the context of distance education: technological, sociocultural, cognitive and management domains. Thus, specifically, PROF_ELE02 tries to plan his teaching activities as well as organizing and structuring his time. His perception of the management domain is noticeable in his absences. PROF_ELE01 literally follows the course syllabus and a specific handout. In this case, there is a need for PROF_ELE01 to expand his mastery of management.

To prepare the *activities*, PROF_ELE01 usually writes the texts and the exercises come from this text and are reflective in nature, with open questions for the student. He reported that students sometimes question the difficulties of finding the questions in the text. The teacher highlighted the remarkable presence and interaction of the tutor in his course.

Every first week of the fortnight, the teacher PROF_ELE01 prepares a complementary reading and from this comes questions with an analytical focus. A kind of "problem-solving". These questions are always open-ended. There are no objective questions. As for the corrections to the exercises, the teacher makes it clear that his form of assessment is to force the learner to study the content. The answers are made available to the students. He doesn't usually give marks for the exercises, but they serve to consolidate the teaching. As for *selecting* the content to teach, the teacher reports that he uses the book indexes themselves to define his selection, as they already contain the sequence. Each fortnight has a key title. Within each application, there are more important subtitles.

For teacher PROF_ELE02, the *preparation of* learning *activities* is done as follows: he follows the initial planning established by the activities, with weekly virtual classes and sending the tasks at the end of the fortnight; however, when the students request more detail on a topic or he notices the doubts and questions of the learners, he holds extra virtual classes and makes the complementary material available.

As for the *preparation and correction of assessments*, PROF_ELE02 reports that he prepares a database with a large number of questions, which can be used in both *online* and written tests at the beginning of the course, with the solutions. After the tests have been taken and made available, the corrections are made. As the questions usually require reasoning and do not have a single solution, it is necessary to consider the logic and technique adopted by the student.

When it comes to selecting content, the same teacher, PROF_ELE02, points out that this item is constantly changing:

> I follow the syllabus, but constantly update the content with new research on the internet, books or magazines. Above all, when I notice students' interest in a particular topic or subject, I try to encourage and motivate them to do research. For example, in the Embedded Electronics subject, topics that are always covered and which arouse great interest among students are robotics and home automation. I try to show them simple examples that they can do and suggest that they do their final work on this subject (Professor PROF_ELE02).

When he reports that he updates his content, it is clear that PROF_ELE02 is concerned with establishing links between the student's experience and technical/scientific knowledge, taking into account the students' profile and learning styles. It not only uses the proposed planning, but updates it. According to Behar (2013), this is a fundamental didactic skill for distance learning teachers.

Professor PROF_ELE01, in relating classroom teaching to distance learning in technical education, raises an important concern:

> I think it's very difficult to have a technical course without practical activities. I think a technical course can be very good if it's the opposite; if it's practically all practical. People are hands-on with equipment, whether it's mechanical, electronic or environmental. They have the physical materials in their hands. The lack of practical activities is a major complicating factor for distance learning. I feel that our distance learning course, at least the one I work on, favors something of an illusion. Teaching technical subjects without practical materials. Without the person doing it. Perhaps to solve this, there could be different materials, more face-to-face meetings. For me, this is a major complicating factor for learning (PROF_ELE01).

For PROF_ELE02, this face-to-face and distance learning relationship is extremely appropriate to the country's reality, as it enables access to technical knowledge in locations where a face-to-face school would be unfeasible, opening up opportunities for people who are unable to travel to a technical center. The world's largest technical institutions are investing heavily in this model, because they see the advantages of the virtual environment, especially in terms of flexible timetables and minimal travel, combined with the reduction in the physical structure required, which is now focused on laboratories for face-to-face practical classes.

5.2.6.3.2 Teaching and Learning in Distance Education - Electronics

When asked about his perception as a teacher in relation to students' expectations, PROF_ELE01 made it clear that most of them enter the course without any conviction, they enter out of mere curiosity. In this case, they give up at the first difficulties. The interpretative and highly reflective readings lead students to give up. The few students who stay resonate well. Both the student and the teacher in distance education must have good study discipline.

PROF_ELE02 reported that, just as in face-to-face courses, those who dedicate themselves properly to learning are effective in distance learning. According to PROF_ELE02, the biggest complaints from students relate to the lack of practical classes or resources at the centers for them to experiment.

When asked about the fact that learning in distance education depends a lot on the student's effort and discipline, and whether learning in distance education meets his expectations as a teacher, PROF_ELE02 replied that:

> Discipline, yes. Students have to set a timetable for studying, doing activities and assignments. But I think that the effort required is the same as in the face-to-face modality, because despite the distance from the teacher, which requires a greater effort to understand, it doesn't require traveling and keeping to fixed timetables. [...] Yes, it meets and often exceeds my expectations, because it requires a lot of creativity and the development of various skills, and this is transformed into great knowledge for the teacher. (Teacher PROF_ELE02).

As for comparing the difficulties encountered by students in face-to-face teaching at with distance learning, PROF_ELE01 says that contact between teacher and student is fundamental because it is a highly technical subject:

> We have to put our finger on it and show it. Look at this, look at that. This verbal communication is much more efficient in person. The fact of putting your finger up and showing, talking, drawing attention, I think it awakens learning. Even if we recorded videos and other materials, the lesson would be cold. The main differences in difficulties between face-to-face and distance learning are related to communication. Holding hands and showing makes the difference. But distance learning is something there's no going back on. It's a trend.

Distance education is possible in more theoretical subjects, for example philosophy, which would be more appropriate (Professor PROF_ELE01).

For PROF_ELE02, the difficulties are also similar. He also points out that the difficulties are even greater in face-to-face education when you consider the lack of structure at the centers, such as the absence of libraries and practical classrooms. PROF_ELE02 made two important points when he was asked whether distance learning actually encourages autonomy and self-direction in students' studies. He stated:

Absolutely, and this is one of my criticisms and suggestions for improvement. I believe that these points should be clearly explained to students at the beginning of the course. Few establish a daily study schedule, most leave the activities and tasks until the end of the fortnight. They need to understand that a flexible timetable requires a lot of discipline and organization (Professor PROF_ELE02).

On the other hand, PROF_ELE02 reports something worrying. When asked if distance learning allows students to understand the relationship between theory and practice, he said: "Not in the current situation. There is a lack of human resources (trained tutors) and structure (equipment, materials, etc.) at the centers to carry out practical activities".

Professor PROF_ELE01 reports that distance education is more difficult than face-to-face education when it comes to reconciling theory with practice on the part of the student. He understands that this is also difficult in face-to-face courses, but in distance learning this is reinforced.

With regard to *monitoring* the construction of student knowledge, the teacher PROF_ELE01 stated that this is basically done through test results and *chat.* Although participation in the *chat*, even though it lasts an average of two hours, is very small. In short, as far as monitoring is concerned, tests and *chats* are the main tools.

For PROF_ELE02, it is a great challenge to teach a subject that integrates and requires students to use the knowledge they have acquired throughout the course. However, he said that he is able to see how the students have evolved over the stages, realizing each one's deficiencies and the shortcomings of the process. He also said that he finds it very difficult for most of them to put together the knowledge they have acquired and understand the applications of what they have been taught. The students know the pieces, but they can't put them together. Most of them simply do or copy the activities for the assessments and to pass the subject. Sometimes they are not led to apply the knowledge they have acquired to practical situations, but only to memorize the concepts. So much so that they expect the tests to be based on closed questions so that they don't reason, just repeat the information in the material.

As for the course's Political Pedagogical Project, the teacher PROF_ELE01 declares that he has no knowledge of it, let alone interaction with this document. Of the resources (video lessons, printed texts, among others) declared in the PPP for use in the course, the teacher makes it clear that he hasn't used the video lessons, even though the students have asked him to. He also doesn't use games, because he doesn't know if it's possible to use games in the subject he teaches. He has tried to use complementary texts, written by himself, making them available virtually. He still uses printed texts from another university.

PROF_ELE02 points out that he needs to devote more time to getting to know the course's PPP. He points out:

This is a point I need to improve and pay more attention to. I have a basic knowledge of the PPP for the Electronics course, and the coordination guidelines. I don't know about the other courses. I believe that publicizing it on the platform and in the teachers' room would help (Professor PROF_ELE02).

PROF_ELE02 also points out that he tries to use most of the resources proposed in the PPP, stating that:

[...] every two weeks we have: video lessons - videos with virtual lessons, video lessons from other sources and practices. Printed support text - containing complementary information to the course content booklet and usually using different language in order to give the student a different approach to the same subject. Texts and complementary activities - The activities and tasks involve practical activities, with texts and support material. Lectures. Every week I give a virtual lesson covering the fortnight's content. I make very little use of the games and activities on the AVA because I don't know how to use them properly. I find the typed *chat* with many students very unproductive, so much so that we abandoned this option and started using weekly videoconferencing (Professor PROF_ELE02).

In this case, PROF_ELE02, even with little knowledge of and interaction with the institutional PPP, proactively seeks alternatives to improve his strategies. When he was shown the pedagogical proposal of the PPP for the course in question and its strategies for the course, the teacher stated that he uses most of them.

In this way, the teacher seems to demonstrate competence in distance education knowledge. According to Behar (2013), this type of competence refers to knowledge of distance education processes, forms of organization, the pedagogical project, didactic-pedagogical methodologies, evaluation processes, among other aspects of academic management.

5.2.6.3.3 Pedagogical Strategies in Distance Learning - Electronics

Among the various pedagogical strategies available on AVA *Moodle*, the teacher PROF_ELE01 has used chat and forum. However, the Pole tutor has only used the forum. He points out that the forum is better than c *hat*.

Teacher PROF_ELE01 has no knowledge of Learning Objects. He is unaware of this strategy. As for simulators, he has suggested and used an *online* simulator. He suggested it and made it available on the AVA, but it was only during face-to-face meetings that the students actually tried it out and approved it. He also hasn't used *webquests*, animations or *wikis*. For teacher PROF_ELE01, the strategies that most contribute to students' autonomous learning are: the forum, c *hat*, video lessons (made by other teachers) and simulators.

Among the strategies used by PROF_ELE02 are: meeting tools (videoconferencing with collaboration) such as *Open Meetings*[1] *and TeamViewer*[2] . These tactics make it possible to hold virtual classes (meetings) and, according to the teacher, the tool has proved to be excellent and has greatly strengthened contact with his students. He also highlighted the use of collaborative mind maps in the development of projects, which facilitated interaction between his students.

Unlike PROF_ELE01's perception, PROF_ELE02 pointed out that the strategies that most promote student learning (depending on the class) in general are: a) video lessons and virtual lessons; b) simulators; c) interactive activities; d) animations. The forum came fifth and external materials sixth in their view.

Regarding the *criteria* for choosing strategies to enhance learning, teacher PROF_ELE01 states something important when he says:

> I think the choice of *chat* and forum is perhaps more a matter of convenience. We have to have a lot of time for planning. Comfort and lack of knowledge make us use only these strategies. To do well in distance learning, we have to spend a lot of time planning. I think it's really down to complacency (Professor PROF_ELE01).

In a different way, PROF_ELE02's view of the criteria for improving learning was: a) ease of use for the student; b) the possibility of collaboration, allowing students to interact with each other, and with the tutor and teacher; c) processing time and speed of response. The teacher added that slow tools in which the student takes a long time to get a response are very poorly accepted, which is why group *chat* was not well received. Lastly, he highlighted the stability of this chosen strategy.

---

1 The *Open Meetings* software can be found at: http://openmeetings.apache.org/. This free software aims to: a) offer videoconferencing, instant messaging, "whiteboarding"; b) collaborative document editing and other *groupware* tools for remote communication and transmission. OpenMeetings is a project of the Apache company (www.apache.org). Accessed on: Aug. 6, 2013

2 The *TeamViewer* software can be found at: http://www.teamviewer.com/pt/. Among its various functions, the following stand out here for use in distance education: online meetings - with up to 25 participants; presentations; training sessions - reducing costs by conducting training online; teamwork - with online collaboration with documents in real time, among other functions. Accessed on August 6, 2013.

As for choosing the best strategy, PROF_ELE02's perception corroborates Behar's (2013) perspective on the concept of technological mastery competence. Technological mastery is defined as informational, digital or virtual competence. It is the type of competence related to digital literacy and autonomy. By digital literacy, Behar (2013) explains that it is the fact that "teachers have a spirit of research, criticism and reflection on the information and tools available on the internet". By autonomy, the same author emphasizes that it is the teacher's potential to make decisions and use technologies effectively to enhance learning.

As a *suggestion* for improvement, PROF_ELE01 highlighted the dedication of the course coordinators. As for e-Tec, he believes that the institution as a whole should be more involved in distance learning, as he himself says:

> It seems that it doesn't matter to the school whether they have this course or not, even though the MEC wants it. We know that many schools have been working with this type of education in Brazil and abroad. I don't think there's any way around it. Distance learning is the future. That's what I feel here at the school today. There is no institutional commitment to the idea of distance education. If there is, it's only because of a few people. Based on this institutional commitment, then we can talk about demands, results, lots of things, performance statistics; because in the current situation, I don't even think the statistics would reflect much of the expenditure that the government is having (Teacher PROF_ELE01).

Suggestions were also put forward by PROF_ELE02. These include: a) restructuring the hubs with equipment, tools and components to enable students to carry out experiments; b) selecting tutors for the hubs who have diversified technical knowledge in the area of expertise, in this case electronics. And before the semester the tutor should be trained in the subjects he or she will have to follow.

Currently, there are few tutors with enough knowledge to help students technically, most just provide bureaucratic support and are very passive.

As for planning, PROF_ELE02 also points out that it is highly necessary to have a timetable, a consistent plan for visits to the centers at the beginning of the semester. This way, the students will know clearly what is going to be covered and will have an idea of what is going to be taught.

In addition, PROF_ELE02, concerned about students dropping out, stated that they should take more responsibility for the course, which is free: He says:

> If they want to give up, they have to pay for the handouts. The fact is that they don't have to pay anything and that's why many, not to say most, think: "I'm going to do the course, it's free, I won't miss anything, if it's too hard, I'll give up, I don't have to pay anything." This was said to me by several students when I asked why so many drop out (Professor PROF_ELE02).

5.3 Discussion of results

A number of insights and recommendations were drawn from the data analysis above. Here are some of them:

5.3.1. For all three courses: Computer Science, Environment and Electronics

When synchronous tools such as *chat are* used as teaching-learning strategies, it is of the utmost importance for teachers to be prepared and oriented so that they can use and manage them well with their students. According to the studies and the perception of difficulties in the interviews, it is necessary to use the strategy based on the definition of specific roles and social protocols for the coordination of a session. And if necessary, analyze and verify the possibilities of using tools external to *AVA-Moodle* for chatting that can provide other techniques for conversation and coordination of the subjects discussed.

The need for a physical infrastructure for the production of video lessons and *podcasts* for teachers and students was noted. This is a fundamental point for our reflection since, in distance education, the availability of these resources for efficiency in the teaching-learning process can become a differential.

The production of video lessons would be a differentiator in today's distance learning context. For this, an analysis of the bandwidth provided by the hubs is essential and urgent. Setting up and managing servers within the centers under the supervision of Belo Horizonte. The video lessons would be recorded on the local server at the hubs and the students and teachers would be able to access these videos. Currently, it takes an average of one hour to watch a 15-minute video. City halls would need to invest in the infrastructure. Therefore, municipalities applying to become hubs should analyze their technological possibilities so as not to compromise the teaching-learning process.

It is important to note that, according to e-Tec's Political Pedagogical Project (2008), the educational infrastructure organized at the educational institution, present at the EaD Center, should be complemented by the technology infrastructure at the centers, consisting of computer labs with Internet access, teaching labs, videoconference rooms and administrative and study spaces that guarantee the student the necessary conditions to carry out the academic activities of the course.

In the interviews, some teachers felt that there was a real need to bring course coordinators and teachers closer together. It is suggested that, as well as bringing teachers closer to their respective coordinators, there should be more frequent meetings (in addition to the biannual meetings) to deal specifically with issues related to the use of the pedagogical strategies of the e-Tec Virtual Learning Environment, with a view to making the most of it and improving student learning. In addition to formal meetings, more constant interaction and occasional monitoring by the e-Tec pedagogical team of the classes and strategies used by teachers within the VLE are suggested.

For the training of teachers starting out on the AVA - *Moodle* platform, we suggest not only face-to-face classes, but also video classes. In this way, teachers who are unable to attend the initial meeting in person can somehow receive the minimum training they need to work as e-Tec teachers via video lessons.

Still on the subject of training, it is recommended to train teachers and tutors in the various distance learning techniques, in a similar way to what is being developed at the beginning of the year. However, this should be done over the course of the year, or before classes start, but with a timescale of more than two weeks.

It was clear from the interviewees in the three courses that, for the current level of distance learning at e-Tec CEFET-MG, a great deal of teacher time is needed to develop and prepare the material and classes, and that the current remuneration is low and not in line with the effort and dedication put in. A feasibility study into the exclusive dedication of teachers at e-Tec CEFET-MG is suggested.

Dropouts are a widely discussed issue in distance education. It is recommended that all three courses establish some kind of restriction for students who drop out of the course. Currently, due to the ease of entry to courses, students generally don't have the necessary commitment and drop out at the first difficulty. There are no restrictions or responsibilities for the student. According to interviews with teachers, the idea of being easy to get into, free and with a title from a major educational institution, in this case CEFET-MG, somehow generates a lack of understanding of what is actually on offer. Some kind of feasibility study is suggested in relation to better student retention. Because the course is free, many don't bother and don't take it seriously.

5.3.2. For the Internet Computing course

In the Internet Computing course, we need to pay attention to the number of VLE strategies used to motivate student learning. The issue is not the quantity of strategies used by the teacher, but the quality of some of them. It's not important to diversify the strategies too much, but to use them well in the context of the subjects.

It is essential that the teacher is aware of the gaps that need to be filled in previous learning, i.e. the prerequisites for the student's current learning. The teacher should not transfer this difficulty to other teachers.

It is suggested that each teacher strives for interdisciplinarity, connecting subjects, especially those that are prerequisites, and this should be done ethically so as not to overlap content in the teaching plans.

5.3.3 For the Environment course

Training courses should focus not only on the use of the AVA platform, but also on the role of the teacher and aspects of their teaching practice within the AVA environment. This will provide more insights and meaningful experiences for teachers starting out as e-Tec teachers.

It is suggested that there should be more dialogue and interaction between teacher and tutor at the hubs. The interviews revealed little dialogue between teacher and tutor. The tutor has direct contact with the student, but the teacher is more distant. The tutor could anticipate and propose new approaches to the teacher. Given this interaction, the teacher could propose and develop materials based on the tutor's operational observations. *Feedback* could be faster between tutor and teacher, thus making the teaching-learning process and student satisfaction more efficient.

As for the e-Tec Political Pedagogical Project, it is suggested that teachers have more contact with this document. In addition to verbal guidance, we also suggest that the coordinator provide written guidance, enabling teachers to gain greater visibility of the pedagogical context and the course, as well as greater support for the quality preparation of their distance learning classes.

Among the observations made in the interviews, there was a certain lack on the part of the teachers as to the real use of pedagogical strategies within the AVA *Moodle.* It is suggested that teachers meet more frequently with the e-Tec technical and pedagogical team (every six months and not just annually), with the specific aim of showing teachers the pedagogical possibilities of the tools available for effective use.

5.3.4 For the Electronics course

Some teachers were concerned about the full and obvious involvement of the CEFET-MG institution in the e-Tec project. In view of this, it is suggested that CEFET-MG leaders become more involved in the e-Tec project itself, making this important project set up by the Federal Government even more viable and mobilizing it.

It is also suggested that the use of other tools for virtual classes be made more accessible, in addition to the *chat* made possible by the VLE. It is necessary to investigate and test strategies for virtual classes external to the VLE, which can add more efficient control and management than the current *chat* used in various contexts in the courses.

In the interviews with the electronics teachers, there was a constant emphasis on the need for considerable financial and human investment in distance education at CEFET-MG. Among these investments, it is suggested that the centers be restructured to provide minimum conditions for practical classes. An unpracticed technical professional is dangerous to the market, and could in some way even denigrate the educational institution that trained the student.

One of the professors made an important point when he said: "Would you entrust the construction of your house to an engineer or master builder who has never made a dough and has only seen videos on the internet of how to do it?" The fact is that companies expect trained technicians to have the minimum theoretical and practical knowledge to carry out their work. That's why there needs to be an emphasis on practical face-to-face training in electronics courses.

In addition, students need to be constantly reminded of the need for self-discipline, organization and commitment. It was clear from the teachers' comments that the flexibility of the virtual environment in terms of study hours means that the vast majority of students don't organize themselves and don't dedicate the time needed to study the subject content.

In the interviews, it was clear that most tutors play the role of interlocutor rather than facilitator for the student. They could conduct practical classes at the hubs, set up study groups and help students develop their work and research. Therefore, the role of tutors at the centers needs to be readjusted.

The teachers' lack of knowledge and interaction with the Political Pedagogical Project was declared unanimous in the interviews with the Electronics course. It is suggested that the coordinators reorient teachers and tutors on the content and importance of the e-Tec PPP. In addition to greater dissemination, interaction and discussion around this very important document.

6. FINAL CONSIDERATIONS

At the end of this research, some conclusions have been drawn from all the data collection and analysis work. Recommendations as well as some future work have been proposed for the possible continuation of this research. This chapter will address these issues.

Based on the research proposed, the object of this study was the pedagogical strategies used in the teaching-learning process in the e-Tec CEFETMG technical courses, with the teachers of the technical courses mentioned in this research as the subjects.

The basic theory, method and technique helped us to understand the subject. Collecting and analyzing the data made it possible to reflect on the strategies used by teachers in e-Tec distance technical education. In order to achieve the objectives set out in the research, we sought to identify the strategies used by technical teachers and their potential for improving the learning process in distance learning.

The majority of the teachers interviewed said that of the pedagogical strategies recommended by the e-Tec PPP (virtual classes, learning objects that will be developed throughout the course, simulators, forums, chat rooms, connections to external materials, interactive activities, virtual tasks (webquest), modelers, animations, collaborative texts (Wiki)), the ones they use the most are: *Chat, Forum, Wiki.*

Careful use of strategies in distance learning plays a key role in enhancing the teaching-learning process. However, as indicated in this study, the coherent use of these strategies by distance learning teachers is influenced by various factors, including administrative, technical and financial support, among other factors embedded in the distance learning modality. Therefore, some conclusions and recommendations are suggested so that the level of effective use of strategies by teachers in distance technical education can be leveraged:

(1) Financial and technological investments in distance education make a difference to the educational process. In the meantime, several interviews revealed a lack of technological infrastructure at the centers, such as Internet with sufficient bandwidth to produce more interactive video classes. For example, in some cases live classes. In this case, there is an urgent need to restructure the servers, broadband connections and computer management involving student access. The process of choosing the City Hall should be reviewed on the basis of its technological infrastructure and on the basis of the proposal in the e-Tec Pedagogical Political Project in order to meet the demand for this modality satisfactorily;

(2) Bringing the coordinators of each course and their teachers closer together. In addition

to the biannual meetings, the analysis of this research revealed a real need for time management on the part of the coordinators, teachers and support staff when it comes to meetings. More frequent meetings of a pedagogical nature are recommended, rather than just course observations. In this case, it would be important to re-evaluate the frequency of meetings between the coordinators and their respective technical course teachers, with a view to examining the use of the pedagogical strategies used in the subjects offered.

(3) With regard to teacher training, it is also recommended that the pedagogical team create specific training video classes for teachers on pedagogical strategies and their potential in e-Tec CEFET. Throughout the year, and not just during the initial teacher training, it is suggested that timetables and discussions be created with teachers about their strategies in their respective courses. What's more, as far as tutors are concerned, according to some reports from teachers, they need to make their role more interactive. Tutors could even teach practical classes at the hubs; set up more work teams; help more proactively with work, projects and research with students. In this way, all the teacher's work would be more focused on preparing materials, teaching strategies and the effective use of Virtual Learning Environment tools.

(4) When analyzing the data on student drop-outs, teachers were visibly concerned. Coordinators, teachers and tutors alike should reinforce to students the need for self-discipline when it comes to their distance learning profile. More specifically, at present, due to the ease of entry to courses, students in general do not have the necessary serious commitment, dropping out at the first difficulty, and there is no restriction or accountability for this student. The natural tendency of students is to neglect aspects of self-discipline and commitment until the end of the course. When the student leaves the course, they could pay a sort of refund, at least for the materials and handouts provided to them. In this case, this dropout should be monitored every six months by the coordinators of each course in order to improve student retention in the e-Tec courses surveyed.

(5) Teacher time and dedication in relation to their current salary was one of the difficulties raised by teachers. In order to develop efficient materials and strategies for the teaching-learning process in e-Tec, an analysis of teachers' time is suggested. Evaluating the possibility of some teachers having exclusive dedication would help and contribute greatly to the creation, construction, analysis and testing of more interactive and up-to-date materials in e-Tec.

(6) Interdisciplinarity in the materials made available in the Virtual Learning Environment (VLE) is fundamental. The data analysis showed that teachers should pay special attention

to the gaps in students' prior learning; the prerequisites for the student to learn the current lesson. It is suggested that special attention, review and reorientation be given by the e-Tec teaching team to the previous content prior to the current lesson being taught. In addition, care should be taken when constructing the curriculum so as not to overlap content from other teachers with similar subjects.

(7) Make the e-Tec Pedagogical Policy (PPP) available *online* to all teachers. In addition, it is recommended that all coordinators constantly encourage observation, interaction and discussion with their teachers and tutors, especially in relation to the methodological proposal of the strategies involved in this document. Also in this political context, we realized the need for real support from CEFET-MG leaders in relation to the e-Tec Project. This involvement would bring not only incentives for teachers, but also greater visibility for e-Tec in the context of CEFET-MG as a whole, as well as greater chances of investment and promising initiatives for e-Tec distance technical courses in the coming years.

(8) The investment highlighted above by the e-Tec teachers is essential, especially when it comes to restructuring the centers. There is an immediate need to restructure the centers, especially for practical classes. Practical classes are essential to enhance technical teaching in this modality, since, in distance education, the moments of practical activities consolidate the knowledge previously acquired in the Virtual Learning Environment (VLE), as well as making the teaching-learning process more real, that is, theory and practice come together, enabling students to exercise their professions in a coherent, broad and quality manner, with a view to the progress of the nation.

(9) Re-evaluate the real effectiveness of the *Chat* tool in terms of: technical feasibility; student *feedback*; collaborative nature; administration and control of the tool. In addition to the strategies used by teachers, it is recommended to study and investigate pedagogical tools and strategies external to the Virtual Learning Environment (VLE), such as videoconferencing tools, *chat* (*freeware*) with more efficient interfaces, configurations and controls, thus enhancing the teaching-learning process in these courses.

(10) FUTURE RESEARCH

As an object of investigation for future research, this study has formulated some concerns and proposals with a view to further verification:

a) What competencies (knowledge, skills and attitudes) are essential for teachers and tutors in distance learning, taking into account the fact that it is public technical education in today's digital environment?

b) What levels of investment are required by the e-Tec Brasil project (MEC) in the distance learning centers and hubs throughout Brazil?

c) What would be and how could existing didactic pedagogical models be applied in the distance modality in the educational context, especially in e-Tec where the focus is on technical education?

d) To investigate a proposal for a didactic-pedagogical model/architecture for e-Tec distance technical education, highlighting the organizational elements (the foundation of the pedagogical proposal); content (instructional materials, learning objects, among others); methodological aspects (strategies, activities, forms of interaction, didactic sequential organization for learning) and technological aspects (VLE functionalities, connections, servers, networks, among others). And in this case, all of this in accordance with the Political Pedagogical Project of the courses highlighted in this research.

e) To study and investigate the feasibility of correlating the concepts of teacher competencies in distance education with a didactic-pedagogical model in favor of improving the teaching-learning process in public technical education in e-Tec Brazil.

f) To study the possible distance learning models of e-Tec CEFET in relation to other federal institutions in Brazil from the point of view of the pedagogical strategies used, thus enabling more accurate, more detailed and more substantial research into the progress and advancement of education, distance learning and, consequently, science.

It is true that the teachers who took part in the research are eager to improve the teaching-learning process, and are willing to learn/train in pedagogical strategies to use in their teaching practices. However, as they pointed out, they need greater financial investment and support from the institution so that they can promote technical training that meets the emerging demands of the digital society.

It is assumed that the questions that guided this research have been answered and these answers should be seen not only as points of arrival, but also as starting points for new research needs, in the direction of advances in distance education, specifically in the use of pedagogical strategies by teachers involved in the distance modality.

Distance learning is a Brazilian reality. A reality in which the Brazilian government, through various initiatives for postgraduate, undergraduate and technical courses, has provided training in places where it was not possible before. However, it is first necessary to consider quality education and not just quantity, and above all to develop methodologies that allow, encourage and train distance learning teachers to use the various pedagogical strategies that can contribute to the teaching-learning process of students who use this modality to obtain the desired training.

In this way, we hope that this research can serve as a basis for reflection and debate, as well as a reference for future studies, contributing to the implementation of pedagogical strategies in distance learning in technical, undergraduate and postgraduate education in the country.

REFERENCES

ABNT NBR 14724:2011. **ASSOCIAÇÀO BRASILEIRA DE NORMAS TÉCNICAS. informaçâo e documentaçâo: trabalhos acadêmicos**: apresentação. Rio de Janeiro: ABNT, 2011. Available at: http://www.abntcatalogo.com.br/norma.aspx?ID=86662. Accessed on: July 12, 2011.

ABBRAGNANO, Nicola. **Dictionary of Philosophy.** 2ª. Ed. Sâo Paulo: Martins Fontes, 1998.

ALMEIDA, M. E. B. Distance Education on the Internet: **Approaches and Contributions of Digital Learning Environments**. In: Educaçâo e Pesquisa. Revista da Faculdade de Educaçâo da USP. Sâo Paulo: v.29, n.2, jul./dez.2003.

ANASTASIOU, Léa das Graças Camargos; ALVES, Leonir Pessate. Teaching strategies. In: ANASTASIOU, Léa das Graças Camargos; ALVES, Leonir Pessate.

(Orgs.). **Teaching processes at university. Assumptions for classroom work strategies**. 3. ed. Joinville: Univille, 2004. p. 67-100.

ALVES, Lynn; BARROS, Daniela; OKADA, Alexandra. **MOODLE: Pedagogical Strategies and Case Studies**. Salvador: EDUNEB, 2009.

ALVES, Joâo Roberto Moreira. **Distance education in Brazil: historical synthesis and perspectives**. Rio de Janeiro: Instituto de Pesquisas em Educaçâo, 1994.

ARANHA, Antônia Vitória S.; CUNHA, Daisy Moreira; LAUDARES, Joâo Bosco (eds.). **Dialogues on work: multidisciplinary perspectives**. Campinas: Papirus, 2005.

ARAÙJO, C. F.; MARQUESI, S. C. **Activities in virtual learning environments: quality parameters**. In: LITTO, F. M.; FORMIGA, M. M. (Orgs.) *Educaçâo a distância:* o estado da arte. Sâo Paulo, SP: Pearson Education do Brasil, 2008.

AUSUBEL, D.P; NOVAK, J.D; HANESIAN, J. **Psicologia Educacional.** RJ: Interamericana, 1980.

AUSUBEL, D.P. **Education Psychology. A cognitive view.** New York. Holt, Rinehart and Wiston, 1968.

AURÉLIO, **O minidicionârio da lingua portuguesa**. 4th Edition. Ed. Melhoramentos. Rio de Janeiro, 2002

BADIA, A; MORENEO. **Psychology of Virtual Education.** São Paulo. Ed. Artmed, 2010.

BAPTISTA, João Manuel Pereira Dias. **Technological education and the new**

programs. Porto: Edições Asa, 1993.

BARROS, Aidil de J. Paes; LEHFELD, Neide A. de Souza. **Research Project: methodological proposals**. Rio de Janeiro: Ed. Vozes, 2009.

BARRETO, Alcyrus Vieira Pinto; HONORATO, Cezar de Freitas. **Manual for survival in the academic jungle**. Rio de Janeiro: Objeto Direto, 1998.

BELUCE, Andrea; OLIVEIRA, Katya. **Teaching and learning strategies in online teaching conditions.** Revista Digital Hipertextus, v. 9, n.6, dec, 2012.

http://www.hipertextus.net/volume9/06-Hipertextus-Vol9-Andrea-Carvalho-Beluce_&_Katya-Luciane-de-Oliveira.pdf. Accessed on: 08 May 2013.

BELLONI, Maria Luzia. **Distance Education.** Sâo Paulo: Ed. Autores Associados, 2009.

BEHAR, Patricia Alejandra. **Competences in Distance Education.** Porto Alegre: Penso, 2013

BERGE, Zane L. **Barriers to online teaching in post-secondary institutions: can policy changes fix it?** The Online Journal of Distance Learning Administration, Carrolton, v.1, n.2. 1998. Available at: <http://www.westga.edu/~distance/Berge12.html> Accessed on: 27 Feb. 2013.

BERGE, Zane. L. Educational Technology, v.35, n.1, p. 22-30. 1995.

BRANDÂO, C. R. (org.) **Pesquisa Participante**. Sâo Paulo: Brasiliense, 1999.

BRAZIL. Decree No. 6.301, of December 12, 2007. Establishes, within the Ministry of Education, the Open Technical School System of Brazil - e-Tec Brasil. Available at:<https://www.planalto.gov.br/ccivil_03/_Ato2007-2010/2007/Decreto/D6301impressao.htm>. Accessed on: June 3, 2012.

BRAZIL. Decree No. 5.622, of December 19, 2005. Available at: <http://www.planalto.gov.br/ccivil_03/_Ato2004-2006/2005/decreto/D5622.htm>. Accessed on: April 21, 2013.

BRAZIL. Decree No. 7.589, of October 26, 2011. Establishes, within the Ministry of Education, the Open Technical School System of Brazil - e-Tec Brasil. Available at: <https://www.planalto.gov.br/ccivil_03/_Ato2011- 2014/2011/Decreto/D7589.htm#art9>. Accessed on: June 3, 2012.

BOURDIEU, Pierre; CHAMBOREDON, Jean-Claude; PASSERON, Jean-Claude. Oficio de sociòlogo: **Metodologia de pesquisa na sociologia**. Petrópolis, RJ: Ed. Vozes, 2004.

BOGDAN, Roberto C.; BIKLEN, Sâri Knopp. **Qualitative research in education**. Trad. Maria Joâo Alvarez, Sara Bahia dos Santos and Telmo Mourinho Baptista. Porto: Porto Editora, 1994.

BRANDÂO, Carlos. **What is education?** Sao Paulo: Brasiliense, 1981.

BZUNECK, J. A. **How to motivate students: practical suggestions**. In: BORUCHOVITCH, E., BZUNECK. J. A.; GUIMARAES, S. E. R. (Orgs.). **Motivating to learn: applications in the educational context**. Petrópolis: Vozes, 2010.

CANDAU, Vera. **Didactics in question.** Petrópolis: Vozes, 1984.

CARVALHO, Alex et al. **Aprendendo Metodologia Cientifica**. Sao Paulo: O Nome da Rosa, 2000.

CARLINE, Alda; TARCIA, Rita Maria. **20% distance learning: now what?** Sao Paulo: Ed. Pearson, 2010.

CASTELLS, M. **A galàxia da internet**: reflexôes sobre a Internet, os negócios e a sociedade. Rio de Janeiro: Jorge Zahar Editora, 2003.

CASTELLS, M. **The network society**. The information age: economy, society and culture. 9ª ed. Sao Paulo: Paz e Terra. 2006.

CERVO, A. L., BERVIAN, P. A. and SILVA, R. da. **Metodologia Cientifica.** 6 ed. Sao Paulo: Pearson Prentice Hall, 2007.

CHIZZOTTI, A. **Pesquisa em Ciências Humanas e Sociais**. 6. ed. Sao Paulo: Cortez, 2003.

COUTINHO, M. T da C. and CUNHA, S. E. da. **The paths of research in the human sciences.** Belo Horizonte: Editora PUC Minas, 2004.

COUTINHO, M. T da C.; MOREIRA, M. **Psicologia da Educaçâo**. 9th ed. Belo Horizonte: Editora Lê, 2004.

COMENIUS, Amos. **Didactica Magna.** SP: Ed. Martins Fontes, 2011.

DAVIS, C.; OLIVEIRA, Z. de M. R. **Psicologia da Educaçâo**. 2a ed. Sao Paulo: Cortez, 1994.

COSTA, José Wilson; OLIVEIRA, Maria Auxiliadora Monteiro de (Org.). **Novas linguagens e novas metodologias : educaçâo e sociabilidade**. Rio de Janeiro: Vozes, 2004.

CORONADO, M. **Competências docentes: ampliantion, enriquecimiento e y consolidacion de la pràtica professional.** Bueno Aires: Novaedu, 2009.

CLARKE, J.H; BIDDLE, A.W. **Teaching critical thinking.** NJ Englewood Cliffs: Prentice-Hall, 1993.

CLARK, C. M., & PETERSON, P. L. **Teachers' thought processes**. In M. C. Wittrock (Ed.), Handbook of research on teaching (ch. 9, pp. 255-296). London: MacMillan Publishing Company. 1986.

CRUZ, Dulce M.; BARCIA, Ricardo. M. **Educaçâo a Distância por Videoconferência**. Tecnologia Educacional, Rio de Janeiro, v. 29, n. 150/151, p. 310, jul./dez. 2000. Available at:< http://penta2.ufrgs.br/edu/videoconferencia/dulcecruz.htm> Accessed on: June 15, 2013.

CUNHA, Maria Isabel da. **Teaching as a complex action: the role of didactics in teacher training.** In: ROMANOWSKI, Joana. Et. al. (Org.).

Local knowledge and universal knowledge: research, didactics and teaching action. Curitiba: Champagnat, 2004. p. 31-42.

D'AVILA, Cristina. **Decipher me or I'll devour you: what can teachers do with textbooks?** Salvador: Ed. Edufba, 2008.

DE ROSSI, V L.S. "**Desafio à escola pùblica: tomar em suas mãos seu próprio destino**". in: Cadernos CEDES/Centro de Estudos Educaçâo e Sociedade. Public Policies and Education. Sâo Paulo: CEDES, 2002, n°. 55.

. **Management of the Political-Pedagogical Project: between hearts and minds**. Sâo Paulo: Moderna, 2004.

DEMBO, M. H. **Applying educational psychology**. 5. ed. New York: Longman Publishing Group. 1994.

DODGE, Bernie. **A Rubric for Evaluating WebQuests.** (1998) Available at <http://webquest.sdsu.edu/webquestrubric.html>. Accessed on: 01 Feb. 2013.

DOWBOR, L. **Education, technology and development.** In BRUNO, Lucio (Org.). **Educaçâo e trabalho no capitalismo contemporàneo.** Sâo Paulo: Atlas, 1996. Pp. 17-39.

DOUGIAMAS, M. & TAYLOR, P. **Improving the effectiveness of tools for Internetbased education**, Teaching and Learning Forum 2000, Curtin University of Technology. Available at http://lsn.curtin.edu.au/tlf/tlf2000/dougiamas.html. Accessed on: June 18, 2013.

DOUGIAMAS, M. & TAYLOR, P. **Interpretive analysis of an internet-based course constructed using a new courseware tool called Moodle. Proceedings of the Higher Education Research and Development Society of Australasia** (HERDSA) 2002 Conference, Perth, Western Australia. Available at http://dougiamas.com/writing/herdsa2002/. Accessed on: June 18, 2013.

DIEHL, Astor Antônio; TATIM, Denise Carvalho. **Research in applied social sciences.** SP: Ed. Pearson, 2004.

e-Tec CEFET-MG. Available at: <http://www.etec.cefetmg.br/site/sobre/etec/>. Accessed on: June 3, 2012.

EPSTEIN, Masters. **Portraits of great teachers.** New York: Basic Books, 1981.

FACHIN, Odilia. **Fundamentals of methodology**. 3.ed. Sâo Paulo: Saraiva, 2001. 210p.

FAVACHO, A.M.; MILL, D. **Functions of technological discourse in contemporary society.** Pro-Proposições, Campinas, 2007.

FERNANDES, Luciane A.; GOMES, José Màrio M. Research reports in the social sciences: Characteristics and research modalities. Contexto, Porto Alegre, v. 1, p. 71-92. 2003. Available at:<http://www.ufrgs.br/necon/04- 4%20Relat%C3%B3rios%20de%20pesquisa%20nas%20ci%C3%AAncias%20sociais%20-%20Luciane%20e%20Jo.pdf> Accessed on: 05 Sep. 2006.

FIDALDO, F. **Diâlogos conceptuais sobre trabalho e educaçâo.** Belo Horizonte: Ed. Puc Minas, 2011.

FREIRE, Paulo. **Pedagogy of autonomy: knowledge necessary for educational practice.** Sâo Paulo: Ed. Paz e Terra, 2011.

GABRIELLI, Katia S., ROZENFELD, Cibele C.de Faria, SOTO, Ucy. **The educational forum in virtual foreign language courses as a tool for interaction: a critical analysis of two experiences**. In: RIBEIRO, Ana Elisa et al. (eds). Language, technology and education. Sâo Paulo: Petrópolis, 2010.

GARCIA, Walter E. **Educaçâo: visâo teòrica e pràtica pedagògica.** Sâo Paulo: Ed. Liber Livro, 2012.

GODOY, Maria Helena Pàdua Coelho; MURICI, Izabella Lana; SA, Rosangela Torres. **How to Improve the Competences of School Staff by Implementing the Training Matrix**. Nova Lima: INDG Tecnologia e Serviço Ltda., 4th edition, 2004.

GIL, Antonio Carlos. **Methodology of Higher Education.** SP: Ed. Atlas, 2012.

GIL, Antonio Carlos. **Didactics of Higher Education.** Atlas, 2011.

GREENO, J. G., COLLINS, A. M. and RESNICK, L. R. **Cognition and learning**. In D. C. Berliner and R. C. Calfree (Eds.) Handbook of Educational Psychology. New York: Macmillan, 1996

GRINSPUN, Mirian, P. S. Zippin (org.). **Technological education: challenges and perspectives**. Sâo Paulo: Cortez, 2001.

GRINSPUN, Mirian, P. S. Zippin (org.). **Technological education: challenges and perspectives**. 3ª. Edition. Sâo Paulo: Cortez, 2009.

GIL, A. C. **Como elaborar projetos de pesquisa.** 4 ed. 11 reprint. Sâo Paulo: Atlas, 2008.

GUTIÉRREZ, F. & PIETRO, D. **Pedagogical Mediation: Alternative Distance Education**. Campinas, Papirus, 1994.

HARVEY, David. **The postmodern condition.** 10.ed. Sao Paulo: Loyola. 2001.

HILLERY, P. **Online chat sessions! Chaos or ...?** In: TCC '99 Papers, 1999.

Available at: http://leahi.kcc.hawaii.edu/org/tcon99/papers/hillery.html. Accessed on April 11, 2013.

HYMAN, R. T (1987). Discussion strategies and tactis. in Vieira, Rui; Vieira, Celina.

Teaching/learning strategies. Horizontes Pedagógicos Collection. Lisbon: Instituto Piaget, 2005.

JUNIOR, Dilermando Piva. **EaD in practice: planning, methods and online education environments.** Sao Paulo: Ed. Campus, 2011.

KENSKI, V. M. **Management and use of media in distance education projects.**

Revista E-Curriculum, Sao Paulo, v. 1, n. 1, dec.-jul. 2005-2006. Available at : http://revistas.pucsp. br/

index.php/curriculum/article/viewFile/3099/2042. Accessed on: March 2013.

LAVILLE, C. and DIONNE, J. **The construction of knowledge:** Handbook of research methodology in the humanities. Porto Alegre: Editora Artes Médicas Sul Ltda. 1999.

LEÃO, L. **A complexidade da Hipermidia in O Labirinto da Hipermidia**. SP: Iluminuras, 1999.

LEFEBVRE A. **Social networks**: the pivot of Internet 2.0. Paris: MM2 Editions. 2005.

LEVY, P. **As Tecnologias das Inteligências: o futuro do pensamento na era da informàtica.** Sao Paulo: Editora 34, 1993.

LEVY, P. **Cibercultura**. Sao Paulo: Editora 34, 1999.

LEVY, P. **What is virtual?** Sao Paulo: Editora 34, 1996.

LIBÂNEO, José Carlos. **Didactics.** Sao Paulo: Ed. Cortez, 1994.

LITWIN, E., **Educaçâo a distância: temas para o debate de uma nova agenda educativa**. Porto Alegre: Artmed, 2001.

LOWMAN, Joseph. **Mastering teaching techniques.** Sao Paulo. Ed. Atlas, 2012.

MALHEIROS, Bruno Taranto. **Research Methodology in Education.** Rio de Janeiro: LTC, 2011.

MALHEIROS, Bruno Taranto. **General Didactics.** Sao Paulo: LTC, 2012.

MAIA, C; MATTAR, J. **ABC da EaD.** Sao Paulo: Ed. Pearson, 2008.

MARCONI, M. de A. e LAKATOS, E. M. **Fundamentos de metodologia cientifica.** 7. ed. Sao Paulo: Atlas, 2010.

MARCONI, M. de A. e LAKATOS, E. M. **Técnicas de Pesquisa**: planejamento e execuçâo de pesquisas, amostragens e técnicas de pesquisas, elabor elaboraçâoçâo, anàlise e interpretaçâo de dados. 6. ed. 2. reimpr. Sâo Paulo: Atlas, 2007.

MARION, José Carlos; MARION, Arnaldo Luis Costa. **Teaching methodologies in the business area.** For administration, management, accounting and MBA courses. Sâo Paulo: Atlas, 2006.

MACHADO, N. J. **Imagens do conhecimento e açâo docente no ensino superior.** Caderno de Pedagogia Universitària - USP. No. 5, June 2008.

MASETTO, Marcos T. **New Technologies and Pedagogical Mediation.** Sâo Paulo. Ed. Papirus, 2010.

MASETTO, Marcos T. **Competência pedagògica do professor universitàrio.** Sâo Paulo: Ed. Summus, 2012.

MATTAR, Joâo. **Tutoring and Interaction in Distance Education.** SP: Cengage learning, 2012.

MATTERLART, A. **History of the information society**. Sâo Paulo: Loyola, 2002.

MAZZOTTI, Alda & GEWANDSZNAJDER, Fernando. **Method in the natural and social**

sciences. São Paulo. Ed. Pioneiro, 2001.

MAZZIONI, S. **The strategies used in the teaching-learning process: Conceptions of accounting students and professors**. In: CONGRESSO USP - CONTROLADORIA E CONTABILIDADE, 9., 2009, Sâo Paulo. Proceedings. Sâo Paulo: 2009. Available at: http://www.congressousp.fipecafi.org/artigos92009/283.pdf. Accessed on: Jul. 2013

MICHEL, Maria Helena. **Methodology and Scientific Research in Social Sciences.** Sâo Paulo: Ed. Atlas, 2009.

MILL, Daniel. **Virtual Teaching.** São Paulo: Ed. Papirus, 2012.

MILL, Daniel. **The challenge of quality interaction in distance education.** Cadernos da Pedagogia Year 02 Volume 02 Number 04 August/December, 2008.

MOODLE. **Documents.** Available at: http://docs.moodle.org. Accessed on: June 18, 2013.

MONTEIRO, B. S.; et al. (2006). **Methodology for developing digital learning objects with a focus on meaningful learning**. XVII Brazilian Symposium on Informatics in Education (SBIE).

MOREIRA, M. A. **Teorias de Aprendizagem**, Editora EPU, 1999.

MOREIRA, M. A. and VEIT, Eliane Angela. **Higher education: theoretical and methodological bases.** Editora EPU, 2010.

MOREIRA, M. A. **The theory of meaningful learning: its implementation in the classroom.** Editora UnB, 2006.

MORESI, Eduardo. **Research methodology**. Stricto sensu postgraduate program in knowledge management and information technology. Brasilia: UCB, 2003.

MOREIRA, Marco and MASINI, Elcie. **Meaningful Learning - David Ausubel's theory**. Sao Paulo: Editora Moraes, 1982.

MOORE, M. and KEARSLEY. **Distance Education: an integrated vision**. Sao Paulo: Cengage Learning, 2008.

MORRIS, Libby V.; XU, Haixia; FINNEGAN Catherine L. **Roles of faculty in teaching asynchronous undergraduate courses.** Journal for Asynchronous Learning Networks, Newburyport , v. 9., n.1, p. 65-82, mar. 2005. Available at: http://sloanconsortium.org/jaln/v9n1/roles-faculty-teaching-asynchronous- undergraduate-courses. Accessed on: Apr. 23, 2013.

NASCIMENTO , F.; CARNIELLI, B.L. **Distance Education in Higher Education.**

Expansion with quality. Educaçâo Temàtica Digital, Campinas, v.9. 2007.

NEaD - e-Tec CEFETMG. Available at: <http://www.etec.cefetmg.br/galerias/arquivos_download/Manual_Etec_Cefet_v1 .pdf>. **Student Handbook - e-Tec. Version 1.0.** Accessed on June 2, 2013.

NISKIER, Arnaldo. **Distance Education: The Technology of Hope**. Sao Paulo, Loyola, 1999.

NUNES, S. d. C.; SANTOS, R. P. D. **Pedagogical analysis of educational portals according to the theory of meaningful learning**. Universidade Federal do Rio Grande do Sul, CINTED-UFRGS - Revista Novas Tecnologias na Educaçâo. 4ed, vol.2, 2006.

OEIRAS, J. Y. Y.; ROCHA, H. V. **A modality of computer-mediated communication and its various interFACES**. In: WORKSHOP ON HUMAN FACTORS IN COMPUTER SYSTEMS, 2000, Gramado. Proceedings. Porto Alegre: Instituto de Informàtica da UFRGS, p. 151-160.

PACHECO, S. B. **Aprendizagem e construçâo do conhecimento nas redes digitais.** Rio de Janeiro: UERJ, 2004 (Doctoral thesis).

PALLOFF, R., & PRATT, K. Online learning communities revisited. The Annual conference on distance teaching and learning, 2005. Available at: http://www. uwex. edu/d isted/conference

/resource_library/proceedings/05_1801.pdf. Accessed on: December 2011.

PEDROSO, Gelta M. J. **Critical success factors in the implementation of distance learning programs via the Internet in community universities**. 2006. 147 p. Thesis (Doctorate in Production Engineering) - Federal University of Santa Catarina, Florianópolis, 2006.

PERRENOUD, P. *et. al.* **Training professional teachers. What strategies? What skills?** 2 ed. Ed. Artmed, 2001.

PETERS, O. **Didâtica do ensino a distância: experiências e estádio da discussão numa visão internacional.** São Leopoldo: UNISINOS, 2003.

PIMENTA, Selma. G; ANASTASIOU, Léa das Graças C. **Docência no Ensino Superior.** Sâo Paulo: Ed. Cortez, 2010.

PIMENTEL, M. G.; SAMPAIO, F. F. **Hiperdiàlogo uma ferramenta de bate-papo para decrease a perda de co-texto**. In: SIMPÒSIO BRASILEIRO DE INFORMATICA NA

EDUCAÇÂO - SBIE'2001 , 2001, Vitória (ES). Proceedings: Vitória, nov. 21-23, 2001. p. 255-266.

PETRUCCI, Valéria Bezzera Cavalcanti; BATISTON, Renato Reis. Teaching strategies and learning assessment in accounting. In: PELEIAS, Ivam Ricardo. (Org.) **Didâtica do ensino da contabilidade**. Sâo Paulo: Saraiva, 2006.

PEREZ, Francisco G. and CASTILHO, Daniel Prieto. **Educational mediation.** Buenos Aires: Ciccus, 1999.

PÉREZ, A. et al.. **Teaching models for a virtual campus. Paper presented at EDUTEC'06 - IX Congreso internacional La educación en entornos virtuales**: calidad y efectividad en e-learning, held from September 19 to 22, 2006.

PORTO, Yeda da Silva. **Pedagogical Mediation in Distance Education: Necessary Skills**. Pelotas: Ed da UFPel, 2009.

PRETI, Oreste. **Distance Education: a mediating and mediated educational practice.** In: PRETI, Oeste. Educaçâo a distância: inicios e indicios de um percurso. Cuiabà: UFMT, 1996. p. 15-56.

Institutional Development Project, PDI. Belo Horizonte: CEFET-MG, 2006.

Institutional Pedagogical Project, PPI. Belo Horizonte: CEFET-MG, 2005.

Political Pedagogical Project, PPP for the Technical Courses in the Environment, Electronics and Internet Computing. Belo Horizonte, CEFET-MG, 2008.

REDE e-Tec BRASIL. Available at: <http://portal.mec.gov.br/index.php?option=com_content&view=article&id=12326&Itemid=665/>. Accessed on: June 3, 2012.

REDE e-Tec BRASIL. Available at:

http://redeetec.mec. gov. br/index. php?option=com_content&view=article&id= 11&Itemid=1>. Accessed on May 8, 2013.

REHEN, Cleunice Matos. **Profile and training of the vocational-technical teacher.** SP. Ed. SENAC, 2009.

SALINAS, J. **Methodological changes with ICT. Didactic strategies and virtual teaching-learning environments**. Bordón 56 (3-4) pp. 469-481, 2004.

SALINAS, J. **Didactic models on university virtual campuses: Teachers' methodological profiles in teaching-learning processes in virtual environments**. IX

Encuentro internacional. Virtual Educa. Zaragoza. 14-18, 2008.

SANTOS, E. **Teacher training and cyberculture: new curricular practices in face-to-face and distance education**. Revista da FAEEBA - Educaçâo e contemporaneidade, Salvador, v. 11, n. 17, p. 113-122, jan./jun., 2002.

SARTORI, A.; ROESLER, J. **Distance higher education. Learning management and the production of printed and online teaching materials**. Tubarâo: Unisul, 2005.

SARTORI, A. S. **Communication management in distance higher education**.

2005. 267 f. Thesis (Doctorate in Communication Sciences) - University of Sâo Paulo, Sâo Paulo, 2005.

SAVIANI, Demerval. **Education from Common Sense to Philosophical Consciousness.** SP: Ed. Cortez, 1980.

SEIDMAN, I. **Technique Isn't Everything, but it is a lot**. In: Interviewing as qualitative research. 3ª. ed. [S.l.]: Teachers College Press, 2005. Chap. 6, p. 6377

SELLTIZ, C. et. al. **Research methods in social relations.** São Paulo: Herder, 1976.

SIEMENS, G. **Connectivism**: A Learning Theory for the Digital Age.elearnspace. December, 2004. Available at:

<http://www.elearnspace.org/Articles/connectivism.htm>. Accessed on: 18/11/2011.

SIEMES, G. & TITTENBERGER, PETER. **Handbook of Emerging Technology for Learning.** Available at: <http://elearnspace.org/Articles/HETL.pdf>. Accessed on: Aug. 3, 2012

SIEMENS G. **A Learning Theory for the Digital Age**, In: International Journal of Instructional Technology & Distance Learning. Jan 2005, Vol 2, N 1, pp3

SIGALÉS, C. **Els factores d'influènciaen l'ùs educatiu d'internet per part del professorat d'educació primària I secundària obligatòria de Catalunya**. Doctoral thesis. Department of Educational and Educational Psychology. University of Barcelona, 2008.

SILVA, Robson Santos da. **Moodle: for authors and tutors.** Sao Paulo: Novatec, 2010.

SILVA, Edna L.; MENEZES, Estera M. **Metodologia da pesquisa e elaboração de dissertaçâo**. 3. ed. Florianópolis: UFSC/PPGEP/LED, 2001. 121p. Available at< http://projetos.inf.ufsc.br/arquivos/Metodologia%20da%20Pesquisa%203a%20edica o.pdf> Accessed on: 15 Apr. 2008.

SILVA, Marco. **Teacher training for online teaching.** Sao Paulo: Ed. Loyola, 2012.

SILVA, E. L. da and MENEZES, E. M. **Metodologia da pesquisa e elaboração de dissertaçâo**. 3. ed. rev. atual. Florianópolis: UFSC Distance Learning Laboratory, 2001.

SILVA, A.C.B. **Pedagogical Project. An instrument for management and change.** Belém: Unama, 2000.

SILVA, M. A. **Do Projeto politico do Banco Mundial ao Projeto Politicopedagògico da escola pùblica brasileira**. Cad. Cedes, Campinas, v.3, n. 61, p.283-301, December, 2003

SMITH, M.; CADIZ, J. J.; BURKHALTER, B. **Conversation trees and threaded chats.** SIGCHI Bulletin, Minneapolis, v. 31, n. 3, p. 21-23, jul, 1999. Available at: http://www.acm.org/sigchi/bulletin/1999.3/morales.pdf. Accessed on June 9, 2012.

TARDIF, Maurice; LESSARD, Claude. **O trabalho docente: elementos para uma teoria da docência como profissâo de interações humanas**. 2. ed. Petrópolis: Vozes, 2005.

TARDIF, Maurice. **Teaching knowledge and professional training.** 10ª. Edition.

Petrópolis: Ed. Vozes, 2010.

THOMPSON, J.B. **Ideologia e cultura moderna: teoria social critica na era dos meios de comunicação de massa**. Petrópolis: Vozes, 1995.

TRIVINOS, Augusto Nibaldo Silva. **Introduction to research in the social sciences: qualitative research in education**. 1st Ed. Sao Paulo: Atlas, 2009

VAHL JÙNIOR, J. C. **Use of interface agents to adapt chats to the distance learning context**. Available at: http://www.teleduc.org.br/artigos/zeh_disser.pdf. Accessed on: June 10, 2013.

VERHAGEN, B. V. P. **Connectivism**: a new learning theory? 11/11/2006.Available at:< http://www.surfspace.nl/nl/Nieuws/Pages/ArchiefoudeSURFsites.aspx>. Accessed on: 29/11/2011

VIEIRA, R. M; VIEIRA, C. **Teaching/learning strategies.** Lisbon: Instituto Piaget, 2005.

YIN, Robert K. - **Case Study Research - Design and Methods**. Sage Publications Inc., USA, 1989.

APPENDICES

APPENDIX A - Authorization for research - e-Tec General Coordinator

APPENDIX B - Authorization for research - e-Tec Course Coordinators

APPENDIX C - Interview script with teachers of distance learning technical courses

APPENDIX A - Authorization for research - e-Tec General Coordinator

Belo Horizonte, February 22, 2013. Dear Prof. Dr. José Wilson da Costa, General Coordinator of e-Tec CEFET-MG

I'm Fabio Neves de Miranda, a student on the Master's course in Technological Education at CEFET-MG. My research project has been approved and I'm ready to start my field and documentary research. As the coordinator of the e-Tec project at CEFET-MG, I would like to request authorization to carry out the research, the general aim of which is to investigate the pedagogical strategies used by teachers in the distance learning modality of e-Tec CEFETMG.

If you are interested, I can send you a copy of the interview and the letter introducing the study to the teachers. I would like to thank you for your attention and kindness in giving me the space to carry out this research at such a prestigious institution as CEFET-MG.

I'm happy to answer any questions you may have.

Sincerely,

Fabio Neves de Miranda - E-mail: fabioNmiranda@gmail.com

APPENDIX B - Authorization for research - e-Tec Course Coordinators

From: Fabio Miranda [maijto:febionmiranda(3)amaiLcom1

Submitted: Monday, April 15, 2013 14:07

To: Coordinators: Adelson de Paula; Demostenes Junior; Francisco Magahaes; Adrhna Tonili

Subject: MASTER'S RESEARCH INTERVIEW

Dear Coordinators Adelson. Demostenes and Francisco,

Thank you for the PPP sent a few days ago as authorized by Prof. Dr. José Wilson.

As a Master's student in Technological Education, I will be researching e-Tec under the guidance of Prof. Dr. G. G.

Dr. Adriana Tonini - Master's Degree in Technological Education.

In the Methodology it will be necessary to INTERVIEW 2 teachers from each e-Tec course (Informatica. Medium

Environment and Electronics).

I would appreciate it if you could indicate 2 Professors, what day they are there (Cefetmg) and email and telephone contacts to carry out this research which is so important for all of us.

Thank you very much.

Att

Fabio Neves de Nftanda

Master's Degree Cefetmg

APPENDIX C - Interview script with e-Tec CEFETMG teachers

FIRST CATEGORY - TRAINING AND TEACHING EXPERIENCE:

1. What is your background (technical, undergraduate or postgraduate)? What experience do you have in distance education? How long have you been working in this area? Have you had any training to work in this modality?

2. Tell us about your experience with eTec CEFETMG's AVA *Moodle* (Virtual Learning Environment). Did you take part in any training program on AVA Moodle to use it at e-Tec CEFETMG?

3. Do you have any knowledge limitations that interfere with your use of the AVA Moodle tools? What would these limitations be?

4. In your experience as a distance learning teacher, how do you carry out the following activities?

a. subject planning;

b. development of learning activities;

c. preparing and correcting assessments;

d. selecting content to teach.

5. How do you perceive the combination of face-to-face and virtual environments (semi-presential mode) in technical education?

SECOND CATEGORY - TEACHING AND LEARNING IN EAD

1. The primary objective of teaching is student learning. As an e-learning teacher, do you realize:

a. that learning in distance education meets students' expectations;

b. that distance learning meets your expectations as a teacher;

c. that learning in distance education depends a lot on the student's effort and discipline;

d. that distance learning students have the same difficulties as a 100% face-to-face course;

e. that distance learning encourages autonomy and self-direction in the organization of their studies;

f. that distance learning allows students to understand the relationship between theory and practice.

2. How have you monitored the construction of the student's knowledge at all stages of the teaching-learning process?

3. What is your knowledge and interaction with the teaching-learning methodological proposal of the PPPs (Political-Pedagogical Project) of the e-Tec CEFETMG courses taught?

4. In the e-Tec PPP, on page 108, it says that for each curricular component there are 12 study units. Each study unit can contain: video lessons, printed support texts, games and activities on the AVA, complementary texts and activities, conference work. Which of these resources have you tried to use to improve the teaching-learning process; which are you not interested in using, either because you don't know about them or for some other reason?

THIRD CATEGORY - PEDAGOGICAL STRATEGIES IN E-LEARNING:

1. In distance learning there are various teaching strategies available for teachers to use, including *chats*, forums, learning objects, simulators, external materials, interactive activities, *webquests*, animations, *wiki* (collaborative texts), among others. In addition to these, what other pedagogical strategies do you know and/or use?

2. Within this set of available strategies, which one(s) do you think actually enable the student to learn more independently?

3. For the technical courses you teach, what resources and strategies have you proposed/used in your classes on AVA Moodle?

4. As tools for interacting with students we have in the AVA: forums, chats, email, among others. What criteria do you consider to be fundamental when choosing these tools, which in fact allow for better results in the teaching-learning process?

Thank you for your time and attention!

Fàbio Miranda - Master's student in Technological Education at CEFET-MG

ANNEXES

ANNEX A - Decree establishing the e-Tec Brazil Systems 2011

Presidency of the Republic

Civil House

Legal Affairs Bureau

DECREE NO. 7.589, OF OCTOBER 26, 2011.

Establishes the e-Tec Brazil Network.

THE PRESIDENT OF THE REPUBLIC, in the use of the powers conferred on her by Article 84, items IV and VI, alinea "a", of the Constitution, and in view of the provisions of Article 80 of Law No. 9.394, of December 20, 1996,

DECREE:

Art. 1 The e-Tec Brasil Network is hereby established within the Ministry of Education with the aim of developing professional and technological education in the form of distance education, expanding and democratizing the supply of and access to free public professional education in the country.

Art. 2 The e-Tec Brasil Network will be constituted through the adhesion of:

I - institutions that are part of the Federal Network of Professional, Scientific and Technological Education;

II - of teaching units of national apprenticeship services offering vocational and technological education courses; and

III - of professional education institutions linked to state education systems.

Art. 3 The objectives of the e-Tec Brasil Network are:

IV - to stimulate the provision of professional and technological education, in the distance modality, in a national network;

V I - expand and democratize the supply of professional and technological education, especially in the interior of the country and on the outskirts of metropolitan areas;

VI I - allow initial and continuing professional training, preferably for students enrolled in and graduating from secondary education, as well as for youth and adult education;

VII - to help young people and adults enter, remain in and complete secondary education;

VIII - allow public education institutions to develop research projects and educational methodologies in distance education in the area of initial and continuing teacher training for professional and technological education;

VI - to promote the development of projects for the production of pedagogical and educational materials for the initial and continuing training of teachers for professional and technological education;

VII - to promote, together with public education institutions, the development of projects for the production of pedagogical and educational materials for vocational and technological education students; and

VIII - allow the development of initial and continuing training courses for teachers, managers and administrative technicians in professional and technological education, in the form of distance education.

Art. 4 The Ministry of Education will set up and implement the e-Tec Brasil Network by means of formal adhesion by interested institutions, expressed in a specific agreement, which will establish the commitments of those involved.

Sole paragraph. The Ministry of Education will regulate the procedures for adhesion, qualification and participation of institutions.

Art. 5 In order to join the e-Tec Brasil Network, interested institutions must set up face-to-face support centers to carry out teaching and administrative activities to support the courses on offer.

§ Paragraph 1 The on-site support centers must have adequate physical space, infrastructure and human resources necessary for the development of the on-site phases of the courses and projects in the e-Tec Brazil Network, including for the attendance of students in on-site school activities provided for in current legislation.

§ 2 The in-person support centers will preferably be located in:

I - municipal, state and Federal District public schools;

II - public institutions offering professional and technological education courses; and

III - teaching units of the national apprenticeship services.

§ 30 The Ministry of Education will set the criteria for accrediting in-person support centers, taking into account their ability to adapt to distance learning.

Art. 6 The Ministry of Education will coordinate the implementation, monitoring, supervision and evaluation of the activities of the e-Tec Brazil Network.

Art. 7 The Ministry of Education will provide technical and financial support for the implementation of e-Tec Brasil activities and will regulate the criteria and procedures for their implementation.

Art. 8 The expenses arising from the implantation and implementation of the e-Tec Brasil Network will be covered by the budget appropriations allocated annually to the Ministry of Education and the National Education Development Fund - FNDE.

Sole paragraph. The Ministry of Education and the National Fund for the Development of Education shall make the selection of professional education courses and programs compatible with existing budget allocations, observing the movement, commitment and payment limits of the budget and financial program, defined by the Ministry of Planning, Budget and Management.

Art. 9 Decree No. 6.301, of December 12, 2007., is hereby revoked

Art. 10 This Decree shall enter into force on the date of its publication.

Brasilia, October 26, 2011; 190th of Independence and 123rd of the Republic.

DILMA ROUSSEFF

Fernando Haddad

This text does not replace the one published in the DOU of 27.10.2011

Source:http://www.planalto.gov.br/ccivil_03/_Ato2011 -2014/2011/Decreto/D7589.htm#art9

MIX
Papier aus verantwortungsvollen Quellen
Paper from responsible sources
FSC® C105338

Printed by Books on Demand GmbH, Norderstedt / Germany